Knowledge Management and Organizational Learning

Volume 10

This series is introduced by the International Association for Knowledge Management (www.IAKM.net) with an aim to offer advanced peer-reviewed reference books to researchers, practitioners and students in the field of knowledge management in organizations. Both discussions of new theories and advances in the field, as well as reviews of the state-of-the art will be featured regularly. Particularly, the books will be open to these contributions: Reviews of the state-of-the art (i.e. syntheses of recent studies on a topic, classifications and discussions of theories, approaches and methods, etc.) that can both serve as a reference and allow opening new horizons Discussions on new theories and methods of scientific research in organisational knowledge management Critical reviews of empirical evidence and empirical validations of theories Contributions that build a bridge between the various disciplines and fields that converge towards knowledge management (i.e.: computer science, cognitive sciences, economics, other management fields, etc.) and propose the development of a common background of notions, concepts and scientific methods Surveys of new practical methods that can inspire practitioners and researchers in their applications of knowledge management methods in companies and public services.

This is a SCOPUS-indexed book series.

More information about this series at http://www.springer.com/series/11850

Malgorzata Zieba

Understanding Knowledge-Intensive Business Services

Identification, Systematization, and Characterization of Knowledge Flows

Springer

Malgorzata Zieba
Department of Management, Faculty
of Management and Economics
Gdańsk University of Technology
Gdańsk, Poland

ISSN 2199-8663 ISSN 2199-8671 (electronic)
Knowledge Management and Organizational Learning
ISBN 978-3-030-75617-8 ISBN 978-3-030-75618-5 (eBook)
https://doi.org/10.1007/978-3-030-75618-5

Author's Habilitation thesis originally published by Gdańsk University of Technology Publishing House in 2018 under the title "Wiedza i jej przepływy w firmach oferujących wiedzochłonne usługi biznesowe".
1st edition: © The Author 2018

This Springer imprint is published by the registered company Springer Nature Switzerland AG.
The registered company address is: Gewerbestrasse 11, 6330 Cham, Switzerland

Introduction

Knowledge and its importance for humanity are nothing new—even in prehistoric times, those who had knowledge of finding food or seeking shelter had a better chance of survival than those who did not. Despite the passage of many thousands of years, the ability to create, store, develop, apply and transmit knowledge is still the key driving force of human life, as well as the basis of the civilization development of mankind. Society develops through learning and passing knowledge to younger generations, as well as embedding knowledge in tangible and intangible objects. At the stage of the industrial revolution, knowledge was mainly contained in technological solutions and manufactured products, and nowadays it is more and more often embedded in intangible solutions that are difficult to assess.

One of the main features of the knowledge-based economy is the growing importance of services in developed countries, especially knowledge-intensive business services (Ciriaci & Palma, 2016). To bring new products and processes to the market, companies nowadays need a broad set of skills and knowledge which they often do not have in their resources and which they have to look for externally. They are helped by companies offering knowledge-intensive business services (KIBS), filling the gaps in the knowledge and skills of their clients while allowing them to optimally use their own resources for other purposes. Therefore, over the past 20 years, researchers and policymakers have shown a growing interest in knowledge-intensive business services, mainly due to the growing number of companies offering such services and the growing demand for them. Moreover, such services significantly contribute to the productivity of the industries they serve (Miles, 2005). The aspects related to the functioning of this type of companies have so far been rarely discussed; the main research perspective covered macroeconomic issues related to the entire sector of knowledge-intensive business services.

The subject of interest of this book is knowledge and its flows in small- and medium-sized enterprises (SMEs) offering knowledge-intensive business services (KIBS). The main goal of the work is to present the theoretical foundations and empirical findings regarding knowledge and knowledge flows in the area of knowledge management, to contribute to the development of management theory in this selected area. The study attempts to comprehensively discuss the phenomenon of knowledge flows taking place both inside small- and medium-sized companies offering knowledge-intensive business services (KIBS SMEs) and between KIBS

SMEs and their broadly understood environment, and to explain the importance of knowledge and its impact on the competitive advantage of these companies.

This topic was taken up by the author for several reasons. First, there are still few publications on knowledge and knowledge management in small- and medium-sized enterprises. This topic is best researched in large, mainly international organizations (e.g. HP, Ernst & Young), which often conduct research themselves to be able to better manage their knowledge. The issues of knowledge and knowledge management in large companies have already been quite thoroughly researched and described, while in small- and medium-sized companies they are still an under-explored topic. Secondly, KIBS SMEs are relatively poorly understood in the area of their operations and management methods, including knowledge management. Research on the importance of knowledge and its particular types and knowledge management in this type of companies is still in the preliminary phase and there are few publications dealing with this subject. Third, for companies in the knowledge-intensive business services sector, the importance of knowledge is crucial, which justifies research into how these companies perceive knowledge and its impact on their competitive position, and how knowledge flows within, and between the environment. The subject of knowledge flows is poorly researched in the literature on the subject, and it should be expected that knowledge flows should be a natural phenomenon in companies offering knowledge-intensive business services. On the one hand, organizations of this type store knowledge in the form of knowledge resources, and on the other hand, they constantly need to gain new knowledge, link it with the possessed knowledge and transform into a constantly improving offer for their clients.

This work combines three important research areas, namely: (1) knowledge as a key resource, (2) knowledge processes (including its flows) and (3) knowledge management (Fig. 1). The discussed topic is up-to-date, and an additional,

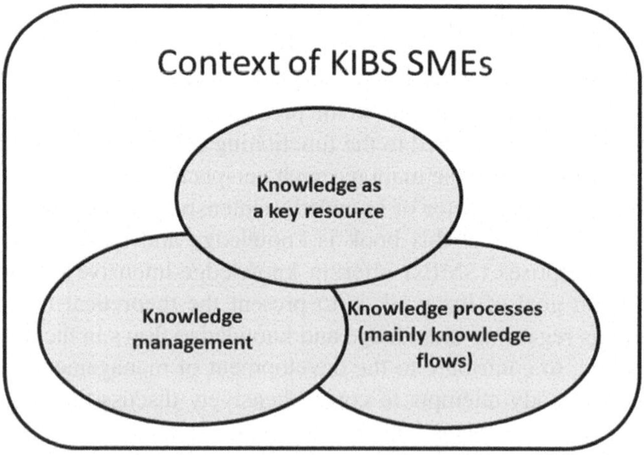

Fig. 1 Research areas discussed in this study. Source: Author's own illustration

significant value of the study is undertaking research in the KIBS sector. Despite a significant increase in interest in KIBS companies and their role in the economy, the number of studies devoted to this issue is still small compared to the number of studies dealing with manufacturing companies (Muller & Doloreux, 2009). An additional justification for the subject matter of this book is the willingness to fill the research gap relating to KIBS companies in the economies referred to as emerging markets, which include the Polish economy. As noted by Muller and Doloreux (2009), out of 82 publications published up to 2005 and analysed in their bibliometric study, only two concerned Central and Eastern Europe. Even though more than a decade has passed since then, and there are studies on emerging markets in Asia (e.g. Liu, 2009; Wong & Singh, 2004), the number of literature on KIBS companies in emerging markets in Central and Eastern Europe is very limited (exceptions include, e.g. Baláž, 2004). Similar conclusions can be drawn on the basis of the analysis conducted by Scarso (2015) in which the author analysed 190 articles on KIBS from 87 journals. Of these studies, only five concerned Eastern Europe. All this confirms that the research in this area is practically unexplored.

Two areas can be distinguished in the book: theoretical and empirical. Theoretical considerations, based on a literature review, made it possible to explain and organize the existing achievements in the area of knowledge management and the sector of knowledge-intensive business services, and thus—to identify a research gap. Particular attention was paid to the aspect of knowledge transfer, which was the starting point for creating the concept of knowledge flows. The effect of theoretical considerations is the concept of knowledge flows, along with its foundations and implications. The developed theoretical concept of knowledge flows, including characteristics of flows, the taxonomy of their dimensions, types of flows, as well as factors determining individual types, were verified in small- and medium-sized companies offering knowledge-intensive business services. The empirical research was aimed at achieving the following research goals:

- diagnosis of the most important types of knowledge used by KIBS companies;
- identifying types of knowledge that affect the competitive position of KIBS companies;
- diagnosis of ways of perceiving and using knowledge by companies offering knowledge-intensive business services;
- identification and characterization of knowledge flows within the organization and between the inside and outside of KIBS;
- diagnosis of potential factors influencing particular types of flows;
- developing a conceptual model for knowledge and its flows in KIBS companies.

The work consists of six chapters, preceded by an introduction and ended with a part containing a discussion of the research results and conclusions and recommendations, as well as indicating the limitations and future directions of research. The first chapter presents selected issues related to knowledge and knowledge management in companies. First, various definitions of knowledge were discussed, and afterward a definition used for the book was proposed. A set of knowledge characteristics and typologies were also developed, and the potential impact of knowledge on the competitive advantage of the organization was explained. This chapter also provides a synthetic overview of the approaches to knowledge management along with their comparison. The presentation of the concept of knowledge management provided a broad background for the concept of knowledge flows.

In the second chapter, the concept of knowledge flow has been presented and defined by the author. Since the concept of knowledge flow is poorly established in the management literature, its theoretical foundations were developed and the taxonomy of knowledge flows was proposed, together with the potential factors influencing the flow of knowledge within the organization and between the organization and its environment. The concept of knowledge flows was created in response to the challenges which contemporary organizations have to face, which include, among others: the need to obtain knowledge "on-demand" and its availability here and now, whenever it is needed; strong connection of knowledge (especially tacit knowledge) with the minds of employees who have it and may, for example, not want to share it; locating knowledge in different places in the organization; high level of formalization or structuring of solutions in the KM area, which are often not available to small- and medium-sized organizations (e.g. due to the costs of their implementation and maintenance); often exaggerated focus on solutions in the area of information technology (IT), which are becoming more and more available, while forgetting about the basic methods of transferring knowledge (e.g. through conversations); the importance of an organizational culture conducive to the exchange of knowledge.

The third chapter presents the definitional issues and the characteristics of knowledge-intensive business services. An attempt was made to organize and define such terms as knowledge-intensive business services, a company offering knowledge-intensive business services or the sector of knowledge-intensive business services. Based on the available literature, the most important aspects related to the definition and classification of KIBS sector services were discussed, and the potential reasons for the growing importance of companies in this sector were analysed. A literature review revealed numerous discrepancies in the definition of these key terms by various KIBS researchers.

The fourth chapter is a prelude to considerations related to knowledge and its flows in companies in this sector. It was also aimed to show a solid basis for undertaking research work. As KIBS companies are dynamically developing in the economy, contributing to the introduction of innovations not only in their own area, but also in the companies of their clients, it is worth examining how they treat their most important resource—knowledge. It should be noted that the vast majority of publications on companies offering knowledge-intensive business services analyse them from the perspective of broadly understood economic sciences, and not management sciences.

The fifth chapter refers to the previous theoretical considerations regarding knowledge, its flows and knowledge management in the context of companies offering knowledge-intensive business services. This chapter also outlines the main research questions concerning, inter alia, knowledge important from the point of view of the functioning of KIBS SMEs, how this knowledge affects the competitive advantage of companies, knowledge flows within KIBS SMEs and between them and the environment, tools and technologies used for knowledge flows or factors determining individual types of knowledge flows. Since these issues have not been studied in the literature so far, and are also characterized by high complexity, it was decided to conduct a case study research, which is an appropriate choice for developing or testing economic theories. To maintain methodological rigour, the procedures for collecting and documenting the data were chosen as recommended (Creswell, 2013, pp. 191–193). The research was conducted in the spirit of the interpretative paradigm, because its goals included extending and verifying the theory of knowledge and its flows in KIBS SMEs, and moreover, in this type of research, the assumption is that "the reality is constructed and cannot be fathomed out or explained with direct reference to universal laws" (Brewerton & Millward, 2001, p. 11). In the case of research on knowledge-related phenomena in the context of KIBS SMEs, many aspects may be perceived as obvious, and at the same time reaching deeper meanings and explaining the reasons for this state of affairs play an important role. Another argument in favour of this paradigm was the author's knowledge of KIBS companies, resulting from previous research conducted in such organizations. In interpretative research, the researcher should be close to the research issue and analyse the research results in a subjective way (Gupta & Awasthy, 2015, p. 12). All these methodological foundations for the study, as well as the study itself, were described in this chapter.

In the sixth chapter, the results of empirical research on knowledge and its flows in companies offering knowledge-intensive business services are presented. The survey covered a total of 14 companies offering knowledge-intensive business services, such as advertising services, accounting services, technical design services and IT services. The obtained empirical material in combination with the available literature allowed for the verification of the theoretical framework regarding knowledge and its flows in the surveyed KIBS companies.

The last part of the book offers the discussion of the most important research results and the presentation of a conceptual model covering knowledge and its flows in KIBS companies. It also presents the limitations of the study and future research avenues.

The reader should be aware that this book concerns a specific sector of reality—the reality in which companies from the KIBS SME sector operate. Although we are currently witnessing a revolution known as Industry 4.0 or the Internet of Things, one should be aware that the above phenomena do not apply to all enterprises to the same extent and not all processes in the economy can be automated. The example of companies from the KIBS SME sector shows that there are still areas that are not suitable for automation and in which people are necessary. Helping and advising other entities, while analysing the entire complexity of their functioning, cannot be addressed with a simple framework.

This work is an improved version of the previous studies. It would not have been written in its present form if it had not been for the support and valuable advice of many people. First of all, I would especially like to thank the Editor of the Book Series, prof. Ettore Bolisani for the possibility to publish this book and all his help at all the stages, as well as substantive and insightful comments on the first version of this book. I would also like to express my gratitude to my colleagues and friends from the International Association for Knowledge Management, especially to prof. Enrico Scarso for joint adventure on KIBS examination and to prof. Constantin Bratianu for his support and useful remarks. Finally, I would like to dedicate this book to my fantastic kids, Ola and Peter, who bring lots of joy and inspiration to my life.

I would like to wish the readers of this book many insightful experiences while reading it and I hope that the book will result in new research areas concerning not only knowledge flows, but also the broadly perceived knowledge-intensive business services sector.

Acknowledgements The author greatly acknowledges the financial support from the National Science Center, Poland, within the grant no. UMO-2016/21/B/HS4/03051, entitled Knowledge management strategies and their determinants in companies from the knowledge-intensive business services sector.

References

Baláž, V. (2004). Knowledge-intensive business services in transition economies. *The Service Industries Journal, 24*(4), 83–100. https://doi.org/10.1080/0264206042000275208

Brewerton, P., & Millward, L. (2001). *Organizational research methods*. Sage Publications.

Ciriaci, D., & Palma, D. (2016). Structural change and blurred sectoral boundaries: Assessing the extent to which knowledge-intensive business services satisfy manufacturing final demand in Western countries. *Economic Systems Research, 28*(1), 55–77. https://doi.org/10.1080/09535314.2015.1101370

Creswell, J. (2013). *Research design. Qualitative, quantitative, and mixed method approaches*. Sage Publications.

Gupta, R. K., & Awasthy, R. (2015). In R. K. Gupta & R. Awasthy (Eds.), *Qualitative research in management. Methods and experiences*. Sage Publications.

Liu, S. (2009). Determinants of service innovative dimensions in knowledge intensive business services: Evidence from PR China. *International Journal of Technology Management, 48*(1), 95–114.

Miles, I. (2005). Knowledge intensive business services: Prospects and policies. *Foresight, 7*(6), 39–63. https://doi.org/10.1108/14636680510630939

Muller, E., & Doloreux, D. (2009). What we should know about knowledge-intensive business services. *Technology in Society, 31*(1), 64–72. https://doi.org/10.1016/j.techsoc.2008.10.001

Scarso, E. (2015). What do we know about KIBS? Results of a systematic literature review. In J. C. Spender, G. Schiuma, & V. Albino (Eds.), *IFKAD 2015: 10th international forum on knowledge asset dynamics: Culture, innovation and entrepreneurship: Connecting the knowledge dots* (pp. 1159–1172).

Wong, P. K., & Singh, A. (2004). The pattern of innovation in the knowledge-intensive business services sector of Singapore. *Singapore Management Review, 26*(1), 21.

Contents

1 Knowledge and Knowledge Management

The concept of knowledge management has become very popular in the last 30 years. This is due to the fact that in the knowledge economy, knowledge has become an important source of competitive advantage, contributing to better profitability and performance of a variety of organizations. Currently, one of the main resources in the economy is not financial or material capital or even human resources, but knowledge itself. Many reasons are contributing to this state, among them the dynamic development of information technology and telecommunications techniques, the growing complexity of the environment in which companies must operate, as well as constant changes and their discontinuous nature.

1.1 Knowledge in Organizations

Before starting a discussion on knowledge management and its flows, it is necessary to discuss what knowledge is and how it can be defined and categorized. In the literature, one can find many definitions and various divisions and classifications of knowledge, such as, for example, the division into explicit and implicit knowledge (Nonaka, 1994) or a division into knowledge about customers, products, processes, as well as about competitors and the external environment (Alavi & Leidner, 2001). In this subsection, various definitions of knowledge, its features and typologies will be presented, as well as the issue of knowledge as a source of competitive advantage.

1.1.1 Knowledge

Knowledge is a complex term that is difficult to define. Definitional problems in the case of the concept of knowledge arise, inter alia, from the fact that, unlike other resources, such as capital or land, knowledge cannot be touched or seen—it is an intangible or even invisible resource. Additionally, one needs to keep in mind that

M. Zieba, *Understanding Knowledge-Intensive Business Services*, Knowledge Management and Organizational Learning 10,
https://doi.org/10.1007/978-3-030-75618-5_1

understanding the concept of knowledge is linked with the metaphor used in defining the concept. By changing the metaphor a person uses, he or she changes completely the conceptual framework and—as a consequence—the definition of knowledge.[1]

Although knowledge can be stored to some extent in documents, the vast majority of it is hidden in people's minds (Evans, 2003, p. 30). From the dawn of time, philosophers have tried to explain what knowledge is (Jashapara, 2011). Plato is considered to be the first philosopher to explore the issues of knowledge, who in his dialogue Theaetetus (306 BCE) stated that knowledge is perception, true judgment supported by justification (Jashapara, 2011, p. 37).

In contemporary management literature, you can find many different types of knowledge definitions that draw attention to its various aspects. For example, according to Davenport and Prusak (1998), knowledge is "a composition of focused experience, values, useful information and professional view, giving the basis for the assessment and assimilation of new experiences and information" (p. 5). According to these authors, knowledge is often stored in documents and repositories, as well as in routines, processes, practices and standards. Importantly, for knowledge to be valuable, it must contain a human aspect—context, experience and interpretation.

Knowledge is what is known or perceived as known (Wellman, 2009). We can base our actions on both false and true knowledge. Knowledge can be contained in the mind of an individual, in the collective consciousness of a group of people, in the cultural norms of a group, or embedded in the processes and methods used to achieve a goal (Wellman, 2009, p. 18). From an organizational perspective, knowledge can be viewed as a "justified belief that enhances an individual's ability to act effectively" (Alavi & Leidner, 2001).

The concept of knowledge often appears in the context of data and information. An important aspect that distinguishes knowledge from, e.g., information or data is the fact that it is born and develops in human minds. Knowledge enables direct action (e.g. making the right decisions, appropriate behaviour, useful work), while information provides meaning and context for such action (e.g. decision criteria, norms of behaviour, detailed job responsibilities) (Nissen, 2006, p. 12). In other words:

- Knowledge enables direct action (e.g. decisions, behaviour, work).
- Information provides meaning and context for actions (e.g. decision-making criteria, behavioural stimuli, working conditions).
- The data answer questions related to the context (e.g.: What is the expected profit in selecting option A? How many accidents at work occurred this year?) (Nissen & Bordetsky, 2011, p. 4).

The following integrative definition of knowledge is proposed:

[1]This has been explained by Bolisani and Bratianu (2018), pp. 7–15.

Knowledge is the entirety of collected, ordered, and processed information supported by reflection, enabling both the individual and the organization to take actions and decisions supported by intuition and wisdom based on the gained experience.

1.1.2 Managerial Features of Knowledge

Knowledge is currently of increasing importance for the operation of organizations and it has several features that make it such an important resource. The first is domination. Knowledge, as a strategic resource of an organization, is more important than other resources. It can be a source of enormous strategic advantage, especially if it meets the conditions of the so-called Barney's VRIO model (Barney, 1997), i.e. it is valuable, rare, difficult to imitate (inimitable) and effectively used (well-organized). The second feature is inexhaustibility. Knowledge is an infinite resource—in the sense that there are no limits to the intensity and scope of its use—and renewable, can be used again and again (McNabb, 2006, p. x). The third feature is simultaneity, which means that knowledge can be used at the same time in different processes and by different people. The fourth feature is non-linearity, which is associated with the lack of a clear correlation between the size of knowledge resources and the related benefits. Sometimes having small but crucial resources of knowledge may result in obtaining a permanent competitive advantage, as opposed to large resources of useless knowledge which constitute a cost for the company (due to the ongoing costs of their maintenance) rather than a source of market advantage.

Another feature is the stickiness of knowledge (von Hippel, 1999). This concept means that it is difficult for knowledge to move from one part of an organization to another. It is related, inter alia, to the frequent localization of knowledge in people's minds, if people move around and share their knowledge, it is less sticky; if not, there is a high level of stickiness. Another feature of knowledge worth mentioning is its liquidity, meaning that knowledge can easily get out of an organization, it can flow out. The uncontrolled leakage of knowledge can sometimes have very negative effects (e.g. loss of competitive position in the market). On the one hand, knowledge can be sticky and it is difficult to pass, for example, from department to department in the company, on the other hand, it can be liquid and "leak" beyond the organization's boundaries. This is a sort of knowledge paradox.

According to Nonaka et al. (2008), knowledge has the following features: (1) subjectivity; (2) process-relativeness; (3) aesthetics; (4) being created through practice. The subjectivity of knowledge is related to the fact that knowledge requires value judgment to be knowledge and this judgment is different for different people performing it. Process-relativeness of knowledge means that knowledge is "created and used in relation with the knowledge of other human beings who exist in relation with others" (Nonaka et al., 2008, p. 10). Aesthetics of knowledge is that knowledge should be judged with regard to what knowledge is being created and what knowledge should be created. In this sense, organizations need to think of knowledge from

Table 1.1 Tacit versus explicit knowledge

Tacit (subjective) knowledge	Explicit (objective) knowledge
Experimental knowledge (body)	Rational knowledge (mind)
Simultaneous knowledge (here and now)	Sequential knowledge (there and then)
Analog knowledge (practice)	Digital knowledge (theory)

Source: Nonaka and Takeuchi (2000), p. 84

the perspective of their values and ideals. Finally, knowledge is being created through practice, when one deals with a particular situation.

In summary, knowledge has many features that distinguish it from other resources that have so far been used by organizations.

1.1.3 Typologies of Knowledge

In addition to the definitions of knowledge that have their origins in different approaches and sciences, and the characteristics of knowledge that influence its characteristics as a resource, the literature also provides an overview of the different types of knowledge and the breakdown of knowledge into subcategories. One of the most popular divisions is the one proposed by Polanyi (1966), who distinguished *tacit* and *focal* knowledge, emphasizing that tacit knowledge is the basis for creating focused knowledge (Stenmark, 2001). This division was developed and clarified by Nonaka and Takeuchi (1995) who divided knowledge into tacit (implicit) knowledge and explicit (available, formal) knowledge. Explicit knowledge can be in the form of words and numbers, therefore it is possible to transfer and store it in an accessible form. Tacit knowledge, on the other hand, is the knowledge that cannot be easily expressed and therefore exists only in people's minds and is expressed through their actions. Knowledge rarely exists in a purely overt or purely hidden form. It is always at least partially hidden in the minds of those who compose it. It is worth emphasizing that from the organizational point of view, explicit knowledge is easier to obtain and store, but at the same time, it is more "diluted" and less powerful or influential than tacit knowledge (Saviotti, 1998).

The differences between tacit and explicit knowledge are presented in Table 1.1.

Based on the division into explicit and implicit knowledge, Nonaka et al. (2000) developed the division of knowledge assets into four groups:

- *Experimental knowledge*—it is tacit knowledge, created in the process of sharing experiences by members of the organization, suppliers and customers.
- *Conceptual knowledge*—it is explicit knowledge, expressed in the form of images, symbols and words (e.g. when creating a product concept or a trademark).
- *Synthetic knowledge*—it is also open knowledge, easy to transfer and share, e.g. knowledge contained in documents, databases, project descriptions, etc.

Table 1.2 Review of the typologies of knowledge according to selected authors—a chronological approach

Author/authors	Knowledge types
Polanyi (1966)	*tacit knowledge*
Lundvall and Johnson (1994) Evans (2003), p. 31	*know-what* *know-why* *know-how* *know-who*
Millar et al. (1997)	Catalogue knowledge—*know-what* Explanatory knowledge—*know-why* Process knowledge—*know-how* Social knowledge—*know-who* Experimental knowledge—*what was*
Nonaka et al. (2000), pp. 20–21	Experimental knowledge Conceptual knowledge Synthetic knowledge Routine knowledge
Gamble and Blackwell (2001)	Personalized, described, embedded knowledge
Alavi and Leidner (2001)	Knowledge about customers Knowledge of products/services Knowledge of the processes Knowledge about competitors and the external environment
Gallouj (2002), p. 258	General and specific knowledge Declaratory, procedural and occasional knowledge Formal knowledge (codified, explicit) and informal knowledge (non-codified, implicit)
Bolisani and Bratianu (2018), pp. 15–19	Rational knowledge Emotional knowledge Spiritual knowledge

- *Routine knowledge*—it is tacit knowledge contained in the daily activities and practices of the organization, organizational routines and culture.

Apart from this division, many other typologies of knowledge can be found in the literature (Table 1.2).

In the literature, one can find, among others, divisions of knowledge based on the area or issue of the given knowledge (e.g. Lundvall & Johnson, 1994; Millar et al., 1997). On the other hand, the division proposed by Bolisani and Bratianu (2018) assumes the separation of rational, emotional and spiritual knowledge,[2] while—as the authors point out—the latter two types of knowledge are very difficult to codify and transmit.

[2]This division is based on the new paradigm of thinking based on a new metaphor, namely "knowledge as energy" and a thermodynamic approach. More details about this approach can be found in Bratianu and Bejinaru (2019a), p. 20 or Bratianu and Bejinaru (2019b), pp. 6–21.

To analyse the knowledge possessed and desired by an organization, the division proposed by Alavi and Leidner (2001) can be used. These authors highlighted four types of knowledge crucial from the organizational point of view. The first type of knowledge is knowledge about your customers. Currently, collecting information and data about customers or their purchasing preferences is insufficient. Organizations need to go one step further and learn about the hidden needs of their customers and the extent to which they are met by the products and services they offer. The second type of knowledge, product knowledge, concerns not only the products offered by a given company but also the products available on the market, their prices, customers who buy them, etc. The third type of knowledge is knowledge about the processes taking place in the organization. It is important to know not only about their course but also about the people involved in them, the possibility of their optimization, etc. The last, fourth type of knowledge is knowledge about the competition and the external environment. With the dynamically changing conditions of the organization's functioning, it is necessary to monitor the external environment and acquire knowledge about the activities of competitive companies, as well as partners, suppliers and institutions potentially influencing the organization. Knowledge about competition and the external environment is necessary to react quickly and effectively to changes and adjust to new operating conditions.

To sum up, it can be stated that knowledge may appear in various forms (e.g. it may be explicit or implicit knowledge), and may also relate to some areas or issues (e.g. knowledge about customers, about processes/services, etc. (Alavi & Leidner, 2001)). Knowledge in organizations is collective, which means that it is not only the sum of the elements of knowledge possessed by individual members of an organization. In this sense, one can define an organization's knowledge base as the collective knowledge that the organization uses for production purposes (Saviotti, 1998).

1.2 Selected Approaches to Knowledge Management

Knowledge management has many definitions, partly due to the huge number and variety of definitions of knowledge presented to some extent in the previous section, and partly due to the multitude of disciplines[3] in which KM is entrenched, and the different approaches to it.[4] Moreover, there are increasingly more voices in the literature that the knowledge of an organization cannot be managed. Despite such

[3]Knowledge management is a relatively young discipline, having its roots in many other areas, such as: information and information systems management, computer science, business management, organization, human resource management or even philosophy. More information on this can be found in the study (Schwartz, 2007).

[4]Representatives of the world of science view KM differently than the world of business. An interesting example is the definition of knowledge management or managing for knowledge, as it is called at Xerox: "creating a thriving work and learning environment that fosters the continuous creation, aggregation, use and re-use of both personal and organizational knowledge in the pursuit of new business value" (Evans, 2003, p. 17).

opinions, knowledge management as a specific discipline within the scientific discipline of "management science" is developing dynamically and there are still many publications on this issue.[5] According to the author of this chapter, knowledge can be managed, just like other intangible resources (e.g. intellectual capital), but it is necessary to focus on specific knowledge resources, processes with their participation and conditions for their course.

Knowledge management is based on the use of resources already owned by an organization, such as well-functioning solutions in the area of information systems, organizational processes or human resource management (Davenport & Prusak, 1998). Skyrme (1997) defines KM as clear and systematic management of key knowledge in an organization and the related processes of creating, collecting, organizing, disseminating, using and exploiting knowledge. As the author emphasizes, KM is associated with the transfer of personal knowledge of the employees of an organization into organizational knowledge that can be shared throughout the organization and properly used. According to uit Beijerse (2000), knowledge management is about achieving the goals of an organization through the effective use of knowledge. This can be achieved by appropriately motivating and stimulating employees and introducing of management support systems and knowledge sharing.

In the literature, there are three main approaches to KM presented: the Japanese approach, the resource approach and the process approach.

1.2.1 Japanese Approach to Knowledge Management

Nonaka and Takeuchi are the authors of the Japanese approach (1995) who created the so-called model of knowledge generation based on the division into explicit and tacit knowledge presented in the previous chapter, the SECI (Socialization, Externalization, Combination, Internalization) model, i.e. the process of creating knowledge through conversion between these two types of knowledge, the concept of ba (the place where knowledge is created and related to) and knowledge assets (Nonaka, 2007). This model is based on the assumption that human knowledge is created and spreads as a result of interactions between implicit and explicit knowledge, and interactions of this kind are called "knowledge conversion" (Nonaka & Takeuchi, 1995, pp. 78–79). According to Nonaka and Takeuchi, the most important thing in an organization is not learning, but knowledge generation, which takes place in the framework of four processes in which tacit knowledge is transformed into explicit knowledge or vice versa. The processes taking place in the knowledge generation/conversion model are:

[5] After entering the term "knowledge management" in the Google Scholar search engine, there are as many as 1,520,000 results with publications that in some way refer to this issue. (23.10.2020).

- *socialization*, i.e. the transformation of tacit knowledge into tacit knowledge—an example is a master–student relationship;
- *externalization*, i.e. the transformation of tacit knowledge into explicit knowledge—an example is the transfer of tacit knowledge possessed by a team into explicit knowledge in the form of metaphors, analogies, hypotheses or models;
- *combination*, that is, transforming explicit knowledge into explicit knowledge—an example is placing documentation on the company's projects in a database;
- *internalization*, i.e. the transformation of explicit knowledge into hidden (tacit) knowledge—an example is the introduction of a product concept into action and practice (Nonaka & Takeuchi, 1995, pp. 78–79).

For example, in the process of socialization, students work with mentors and learn the craft not through language, but observation, imitation and practice. In business conditions, the same principle is followed by on-the-job training. The combination process is based on the use of social processes to combine various resources of explicit knowledge held by individual employees. Employees exchange and combine knowledge through exchange mechanisms such as meetings and telephone calls (Nonaka, 1994). Externalization is very important for the organization because, as a result tacit knowledge possessed by employees becomes open knowledge and available to other members of the organization. In turn, internalization, which consists of transforming explicit knowledge into tacit knowledge, resembles the traditional concept of "learning".

Knowledge conversion processes are seen mainly through the prism of the organization's interior. The SECI model is based on four types of processes: socialization, externalization, combination and internalization, which relate to the situation inside the company. Among the methods of managing knowledge conversion processes, there are several techniques suitable for the transfer of individual types of knowledge (explicit vs. implicit). The methods of managing knowledge conversion processes are presented in Table 1.3.

1.2.2 Resource Approach to Knowledge Management

The resource approach was initiated by Leonard-Barton from Harvard Business School, who created a resource model based on "wellsprings of knowledge" such as key competencies and key skills (Leonard-Barton, 1995). The resource model of knowledge management is rooted in the resource approach developed as part of strategic management. This approach treats knowledge as the company's most important asset, both inside and outside the organization. The adoption of the resource approach in managing organizations and focusing on intangible assets of the enterprise, which are a potential source of competitive advantage, contributed to the increased awareness of the importance of knowledge as a strategic resource. The condition for effective knowledge management is the coexistence of five elements: importing knowledge from the external environment, implementation and integration of new tools and technologies, experimentation, joint problem-solving and key

Table 1.3 Methods of managing knowledge conversion processes

Method	Process	From to	Techniques
Socialization	Implicit knowledge–implicit knowledge	Man–man	Delphi Brainstorm Situational approach IP SSM Structuring of SAST problems Trial and error Learning by doing Comparative analysis Analogies Simulations Consulting Practice
Externalization	Implicit knowledge–explicit knowledge	Man–man Man–information carrier	Delphi Brainstorm Situational approach IP SSM Structuring of SAST problems Trial and error Learning by doing Comparative analysis Simulation IT
Combination	Explicit knowledge–explicit knowledge	Man–man Man–information carrier Information carrier–information carrier Information carrier–human	System analysis Systems engineering Discovering knowledge about databases Synthesis: from qualitative to quantitative approach Comparing systems Collecting data
Internalization	Explicit knowledge–implicit knowledge	Man–man Information carrier–human	Simulation Learning by doing Trial and error Analogies Logical analysis Consulting Practice

Source: Gao et al. (2002)

skills, consisting of physical and technical systems, management systems, knowledge and skills of employees, standards and values (Leonard-Barton, 1995).

The above-mentioned elements of the model concern both the inside of the organization (implementation and integration of new tools and technologies) and its environment (import of knowledge from the environment) and are placed in the present (joint problem solving) and look ahead (experimenting).

1.2.3 Process Approach to Knowledge Management

The process approach is derived from Murray and Myers, who defined the knowledge management process as "all the processes of creating, disseminating and using knowledge to achieve the goals of the organization" (Murray & Myers, 1997). There are many descriptions of knowledge management as a process in the literature. Jashapara defines KM as "an effective learning process related to the search, use and dissemination of knowledge (explicit and hidden), using appropriate technologies and cultural environment, the purpose of which is to increase intellectual capital and organizational efficiency"(Jashapara, 2006, p. 28).

Knowledge management process can be defined as a specific architecture used to acquire and use knowledge resources. The purpose of the KM process should be to facilitate the acquisition and use of these resources for the significant benefit of the organization.[6] When one wants to distinguish between KM process and knowledge-related process, the knowledge management process is a broader, more holistic and strategically oriented term, while the knowledge-related process may refer to a specific action, measure or activity related to knowledge, e.g. acquiring, developing or sharing knowledge in an organization.

The knowledge management process consists of various knowledge-related processes, bringing them together. It is an integral part of the management process and does not replace classic management. The distinction is that classical management focuses on material resources, and knowledge management—on non-material resources (mainly knowledge).

Knowledge-related processes support the exchange of knowledge between organizational units and business processes. An example is processes that support the collection, organization, storage and dissemination of knowledge as a result of business processes or processes that are responsible for assigning skills and expertise to business processes or projects. Knowledge management processes require planning, organizing, motivating, controlling and making decisions, thus managing the organizational knowledge base. In the knowledge management process, activities related to the life cycle of the knowledge are combined in processes related to knowledge. These processes are connected with other processes related to knowledge, as well as with business processes (Maier & Remus, 2003).

The process approach is based on experience and practical solutions used in large consulting companies, such as IBM Consulting Group, Pricewaterhouse Coopers or Ernst & Young. The main promoters of this approach are Davenport and Prusak, who in their book "Working knowledge" focused precisely on the description of

[6]Knowledge Management Process in NASA, https://km.nasa.gov/wp-content/uploads/sites/3/2015/11/Knowledge-Management-Process.pdf (accessed 08/16/2017).

practical experiences in the field of knowledge management in such companies (Davenport & Prusak, 1998). On this basis, they distinguished three main processes in the knowledge management process: the knowledge creation process, the knowledge codification process and the knowledge transfer process. The group of researchers disseminating this trend also includes Probst, Romhardt and Raub (Probst et al., 2000) and Bukowitz and Williams (Bukowitz & Williams, 2000). The first group of researchers identified six knowledge-related processes:

- *locating knowledge*—it is a process aimed at identifying the sources of knowledge, so that members of the organization can easily orient themselves in the internal and external knowledge environment;
- *knowledge acquisition*—it is a process of acquiring knowledge, either from within the organization or from its environment; in the case of acquiring knowledge from outside the boundaries of the organization, there is a risk of rejection of the acquired knowledge by its members;
- *developing knowledge*—it is a process oriented at creating new knowledge within the organization, aimed at creating new competencies;
- *knowledge sharing and dissemination*—it is a process aimed at disseminating knowledge in the organization and providing it wherever it is needed;
- *knowledge exploitation*—it is a process of applying knowledge in an organization, important in that the mere possession or dissemination of knowledge is not enough, the ability to use it is of key importance;
- *preservation of knowledge*—it is a process oriented towards the formalization of explicit knowledge, related to the creation of organizational memory and the selection of knowledge that should be preserved in the organization; includes three stages: selecting knowledge (people, events, processes) worth preserving, giving this knowledge an appropriate form in which it can be stored and updating organizational memory (Probst et al., 2000).

In addition, there are many other typologies of knowledge-related processes in the literature (Table 1.4).

As can be seen (Table 1.4), the Japanese approach is a kind of process approach, because in the model proposed by Nonaka and Takeuchi (1995) there are four knowledge processes: socialization, internalization, externalization and combination.

Moreover, the typologies of knowledge-related processes show mutual similarities and some processes are common to many typologies. These include: acquiring and creating knowledge, storing/organizing knowledge, using/applying knowledge, disseminating/transferring knowledge distribution and using/using knowledge. It is interesting to note the need to remove knowledge that is redundant to the organization or has already become obsolete. Wijnhoven also points to unlearning as one of the processes related to knowledge (Wijnhoven, 2006, p. 10). Research shows that also unlearning is an important part of knowledge management in organizations (Cegarra-Navarro et al., 2011).

Table 1.4 Typologies of knowledge-related processes

Author	Knowledge-related processes
Hedlund (1994)	Gaining of knowledge Knowledge storage Knowledge transfer Applying knowledge Knowledge protection
Skyrme (1997)	Knowledge creation Accumulation of knowledge Organizing knowledge Knowledge dissemination Use of knowledge Continuous exploitation of knowledge
Nonaka and Takeuchi (1995)	Socialization Internalization Externalization Combination
Quintas et al. (1997)	Knowledge creation Gaining of knowledge Capturing knowledge Sharing knowledge Use of knowledge
Jashapara (2006), p. 15	Discovering knowledge Knowledge creation Valuing knowledge Spreading knowledge Use of knowledge
Wijnhoven (2006)	Developing knowledge Maintaining knowledge Knowledge storage Knowledge dissemination Knowledge erasure and unlearning
Fong and Choi (2009)	Knowledge acquisition Knowledge creation Knowledge storage Distribution of knowledge Use of knowledge Maintaining knowledge

Source: own study, based on the mentioned studies

1.2.4 Comparison of Approaches to Knowledge Management

Each of the approaches presented in the above sections has its own characteristics, advantages and disadvantages (Table 1.5). The process approach based on research carried out in organizations managing their knowledge on a daily basis can be considered the most pragmatic. Although it lacks a solid methodological framework, presenting the method of reaching individual solutions, it can be used in many different types of companies. Japanese and resource approaches are a bit more difficult to apply without proper knowledge management preparation, but they pay

Table 1.5 Comparison of the Japanese, resource and process approach to knowledge management

Name of the approach	Japanese approach	Resource approach	Process approach
Characteristics	It is based on the distinction between explicit and implicit knowledge. It is based on the assumption that human knowledge is generated and transferred as a result of social interactions between the two types of knowledge. The common denominator with the process approach is the existence of knowledge-related processes (knowledge conversion processes).	It combines elements of the internal structure of the organization (implementation and integration of knowledge) with elements of the environment (importing knowledge) in the present and future dimensions. It comes from traditional strategic management.	It describes the various stages of the knowledge management process in the organization.
Disadvantages	The measures described are difficult to apply, as is measuring their effectiveness. The approach is rooted in the culture of the Far East, in large enterprises with a complex structure, which may cause difficulties in its implementation on other grounds (e.g. in other countries or types of enterprises).	The approach is too deeply rooted in the realities of the industrial economy. Organizations mainly focus on short-term activities rather than looking ahead. The approach follows the traditional concept of strategic thinking, it is not very innovative.	This approach lacks clear guidelines on the methodology for developing appropriate solutions.
Advantages	It emphasizes the value of knowledge possessed by employees of the organization. It is characterized by an extensive philosophical sphere. It is a holistic concept (includes a process approach—SECI model—and resource approach—knowledge assets and ba).	It emphasizes the role of employees and their knowledge. Attention is directed to the aspect of organizational culture and cooperation/joint problem solving.	It is firmly established in practice (created on the basis of the experience of large consulting companies). Can be used in a wide range of enterprises.

attention to other important aspects such as key employee competencies (resource approach) or transforming and creating new knowledge thanks to employees (Japanese approach). It can also be concluded that these approaches are complementary.

For example, Jashapara (2011) proposed an integrated model of knowledge management, in which he combined various aspects of KM. According to him, knowledge management covers four aspects: organizational learning, culture, systems and technologies and strategy. Organizational learning pays attention to the human resource perspective that enables organizations to seek, use and disseminate knowledge. Information systems and technologies, in turn, are seen by some as a tool for transforming an organization into a knowledge organization. Culture is another important element of KM, because only openness to the environment, the atmosphere of cooperation, the lack of fierce competition between employees and fostering innovative, extraordinary solutions allow for free flow of knowledge between employees and the environment of the organization. The strategy, on the other hand, provides KM a key place in the organization.

In the classic approaches to knowledge management outlined above, the centre of the organization is central. The researchers' attention is focused mainly on internal resources, which should be well-organized so that they are easily accessible and usable. The aforementioned researchers indeed point to the need to be open to the environment and obtain knowledge from the outside, but this is not the main element of these approaches.

At the same time, in the era of intense network relations, both formal and informal, the analysis of approaches to knowledge management should also focus on inter-organizational relations. Organizations nowadays are rarely able to function without close cooperation, e.g. with their customers, suppliers, research and development centres or even competitors.

Among the various groups creating network relations with the enterprise, one of the most important (and often the most important) is customers. It is their knowledge that is a very important element supporting the operation of the organization. Without the knowledge of customers, as well as knowledge of customers and their needs, it is difficult to achieve success in the market. Therefore, the concept of customer knowledge management will be described in the next section.

1.2.5 Customer Knowledge Management

In the context of various approaches to knowledge management, it is worth discussing the concept of customer knowledge management (CKM). Thanks to customer knowledge management, companies can more easily identify emerging market opportunities, thus ahead of their competitors, constructively change the well-established way of operating, and more efficiently create economic value not only for themselves, but also for their shareholders and—above all—customers (Gibbert et al., 2002). Importantly, in the case of CKM, knowledge is not the property of the company, but other entities (individual and organizational), which may be willing to share their knowledge or, on the contrary, not feel like it. It is these other entities (various types of stakeholders: customers, suppliers, partners or competitors) that are involved in the knowledge creation process in CKM—together with the organization, they co-create knowledge to create value for both parties (Paquette, 2011). Therefore, the flow of knowledge between the internal and external environment of the organization plays an important role in the CKM concept.

The CKM concept goes beyond customer relationship management (CRM)[7] and differs from it in many different ways. When it comes to the direction of interactions, in the case of CRM they take place one way (from the client to the company), while in the CKM—both ways (from the client to the company and from the company to the client). These concepts also differ in terms of the medium used—in CRM they are mainly technological tools, while in CKM—a personal relationship. The subject of interest is also different—for CRM it is data, and for CKM—customer experience. Another difference relates to the purpose of applying individual concepts: in CRM, the goal is to identify profitable customers, and in the case of CKM, the goals include collecting customer ideas, identifying areas for service improvement or developing a new product. Also, the role of workers in these concepts is different (Murillo & Annabi, 2002).

The creators of the CKM concept defined five of its different styles. These include: prosumption, collaborative learning, team-based, mutual innovation, creative communities and shared intellectual property (Gibbert et al., 2002). Prosumption, meaning the client's dual role: producer and consumer, is a term proposed in 1980 by Toffler (Toffler, 1980). The idea of prosumption generally concerns the transfer to the client of some of the tasks that were previously performed by the organization.[8] Collaborative learning, based on teams, is a way of exchanging knowledge between an organization and its customers. It can take a real dimension (e.g. actual customer meetings with representatives of the organization and discussions about products/services), as well as a virtual one (e.g. through moderated by companies Internet portals where customers present their experiences with products/services). The idea of mutual creation of innovations was proposed for the first time by von Hippel, who stated that most product innovations in some areas are created by end users of these products, who have specific knowledge about the products and know their needs (von Hippel, 1988, pp. 4–5). Mutual innovations are undertakings based on mutual cooperation between clients and the company in innovative processes. On the one hand, they allow for the sharing of the costs of such innovations, and on the other, they may involve the risk of leakage of valuable

[7]CRM (Customer relationship management) is "the combination of practices, strategies and technologies that companies use to manage and analyse customer interactions and data throughout the customer lifecycle. The goal is to improve customer service relationships and assist in customer retention and drive sales growth". Source: https://searchcustomerexperience.techtarget.com/definition/CRM-customer-relationship-management, date of access: 15.11.2020.

[8]An example of prosumption is that IKEA sells not so much a finished product, but its components together with an instruction based on which the consumer can assemble it. Another example is the introduction of online banking, in which the customer independently performs operations on his account, it does not have to be done by the cashier at the bank's office.

information and knowledge. Creative communities are when a company organizes its clients into groups with similar expertise and encourages them to interact to generate new knowledge. These groups are characterized by long-term cooperation, the same interests and the willingness to create and share valuable knowledge. Unlike traditional communities of practice, these groups cross the boundaries of an organization and create value for many different organizations (Paquette, 2011). The last style, shared intellectual property, involves very intensive cooperation between the organization and its clients, resulting in even partial assumption of the intellectual property rights of the organization by its clients. In this style, there is more than a joint creation of products or services by the organization and its customers: the company and its customers contribute to the future image of the company, in which its success becomes the success of customers and vice versa (Gibbert et al., 2002).

1.3 Summary

To sum up, this chapter has made the reader acquainted with some key terms and concepts. First, it presented the complexity of knowledge and the reasons for the definitional problems. Based on the analysed definitions, the author has proposed the following, integrative definition of knowledge:

Knowledge is the entirety of collected, ordered, and processed information supported by reflection, enabling both the individual and the organization to take actions and decisions supported by intuition and wisdom based on the gained experience.

This definition is necessary to set a certain mental paradigm to a better understanding of knowledge in an organization and the concept of knowledge management. Second, several managerial knowledge features have been described, such as: knowledge domination, inexhaustibility, simultaneity, non-linearity, the stickiness of knowledge, subjectivity, process-relativeness, aesthetics and being created through practice. All those features make knowledge a special resource in the organization, far different from the other ones (e.g. labour or natural resources). Third, the plethora of knowledge typologies have been presented in a chronological order, starting from tacit and explicit knowledge (Nonaka & Takeuchi, 1995), and finishing with the rational, emotional and spiritual knowledge (Bolisani & Bratianu, 2018). Some of those typologies are especially useful for the organizations to examine their knowledge, e.g. the typology dividing knowledge into: knowledge about customers, knowledge of products/services, knowledge of the processes, knowledge about competitors and the external environment (Alavi & Leidner, 2001). It also needs to be remembered that knowledge in organizations is collective, which means that it is not only the sum of the elements of knowledge possessed by individual members of an organization, but much more.

Fourth, the concept of knowledge management has been analysed and the following conclusions can be drawn from this analysis.

1. Knowledge management is often seen through the prism of the activities with which it is related, such as the implementation of knowledge processes (e.g. acquiring, collecting, disseminating, etc.).
2. Knowledge management is supposed to contribute to the achievement of a company's goals and to achieve better results than in the absence of knowledge management.
3. Knowledge management should refer to knowledge that is useful from the point of view of an organization—in other words, an organization should identify knowledge relevant to its activities (especially knowledge constituting a strategic resource) and manage it effectively.
4. Knowledge management has a holistic dimension in the sense that it concerns the entire organization and should be implemented systemically and continuously.

Fifth, three main approaches towards KM (the Japanese approach, the resource approach and the process approach to knowledge management) have been analysed from the perspective of their characteristics, advantages and disadvantages. These approaches are different, however, they are complementary to each other. Additionally, it is important to remember that regardless of the KM solution that the organization decides to implement, it is important that knowledge management contributed to the better results of an organization. For this purpose, it has to be comprehensive, supported by leadership, embedded in organizational culture, benefited from technologies, but always placed the human in the centre.

References

Alavi, M., & Leidner, D. E. (2001). Knowledge management and knowledge systems: Conceptual foundations and research issue. *MIS Quarterly, 25*(1), 107–136.

Barney, J. B. (1997). *Gaining and sustaining competitive advantage.* Addison-Wesley Reading, MA.

Bolisani, E., & Bratianu, C. (2018). Emergent knowledge strategies. Strategic thinking in knowledge management. In Knowledge management and organizational learning (Vol. 4, 1). Springer. doi: https://doi.org/10.1007/978-3-319-60657-6

Bratianu, C., & Bejinaru, R. (2019a). The theory of knowledge fields: A thermodynamics approach. *Systems, 7*(2), 20. https://doi.org/10.3390/systems7020020

Bratianu, C., & Bejinaru, R. (2019b). Knowledge dynamics: A thermodynamics approach. *Kybernetes, 49*(1), 6–21. https://doi.org/10.1108/K-02-2019-0122

Bukowitz, W. R., & Williams, R. L. (2000). *The knowledge management fieldbook.* Financial Times/Prentice Hall.

Cegarra-Navarro, J. G., Sánchez-Vidal, M. E., & Cegarra-Leiva, D. (2011). Balancing exploration and exploitation of knowledge through an unlearning context: An empirical investigation in SMEs. *Management Decision, 49*(7), 1099–1119. https://doi.org/10.1108/00251741111151163

Davenport, T. H., & Prusak, L. (1998). *Working knowledge: How organizations manage what they know.* Harvard Business Press.

Evans, C. (2003). *Managing for knowledge: HR's strategic role.* Routledge.

Fong, P. S. W., & Choi, S. K. Y. (2009). The processes of knowledge management in professional services firms in the construction industry: A critical assessment of both theory and practice.

Journal of Knowledge Management, 13(2), 110–126. https://doi.org/10.1108/13673270910942736

Gallouj, F. (2002). Knowledge-intensive business services: Processing knowledge and producing innovation. In J. Gadrey & F. Gallouj (Eds.), *Productivity, innovation and knowledge in services* (pp. 256–284). Edward Elgar.

Gamble, P. R., & Blackwell, J. (2001). *Knowledge management: A state of the art guide*. Kogan Page Publishers.

Gao, F., Li, M., & Nakamori, Y. (2002). Systems thinking on knowledge and its management: Systems methodology for knowledge management. *Journal of Knowledge Management, 6*(1), 7–17. https://doi.org/10.1108/13673270210417646

Gibbert, M., Leibold, M., & Probst, G. (2002). Five styles of customer knowledge management, and how smart companies use them to create value. *European Management Journal, 20*(5), 459–469. https://doi.org/10.1016/S0263-2373(02)00101-9

Hedlund, G. (1994). A model of knowledge management and the N-form corporation. *Strategic Management Journal, 15*, 73–90.

Jashapara, A. (2006). *Zarządzanie wiedzą: zintegrowane podejście*. Polskie Wydawnictwo Ekonomiczne.

Jashapara, A. (2011). *Knowledge management. An integrated approach*. Pearson Education.

Leonard-Barton, D. (1995). *Wellsprings of knowledge: Building and sustaining the sources of innovation*. Harvard Business School Press.

Lundvall, B. Å., & Johnson, B. (1994). The learning economy. *Journal of Industry Studies, 1*(2), 23–42. https://doi.org/10.1080/13662719400000002

Maier, R., & Remus, U. (2003). Implementing process-oriented knowledge management strategies. *Journal of Knowledge Management, 7*, 62–74. https://doi.org/10.1108/13673270310492958

McNabb, D. E. (2006). *Knowledge management in the public sector: A blueprint for innovation in government*. ME Sharpe.

Millar, J., Demaid, A., & Quintas, P. (1997). Trans-organizational innovation: A framework for research. *Technology Analysis & Strategic Management, 9*(4), 399–418. https://doi.org/10.1080/09537329708524294

Murillo, M. G., & Annabi, H. (2002). Customer knowledge management. *Journal of the Operational Research Society, 53*(8), 875–884. https://doi.org/10.1057/palgrave.jors.2601365

Murray, P., & Myers, A. (1997). The facts about knowledge. *Information Strategy, 2*(7), 29–33.

Nissen, M. E. (2006). *Harnessing knowledge dynamics: Principled organizational knowing & learning*. IRM Press.

Nissen, M. E., & Bordetsky, A. (2011). Leveraging mobile network technologies to accelerate tacit knowledge flows across organisations and distances. In G. Trentin (Ed.), *Technology and knowledge flows: The power of networks* (pp. 1–25).

Nonaka, I. (1994). A dynamic theory of organizational knowledge creation. *Organization Science, 5*(1), 14–37. https://doi.org/10.1287/orsc.5.1.14

Nonaka, I. (2007). *The knowledge-creating company*. Harvard Business Review, July–August.

Nonaka, I., & Takeuchi, H. (1995). *The knowledge-creating company: How Japanese companies create the dynamics of innovation*. Oxford University Press. https://books.google.pl/books?id=B-qxrPaU1-MC

Nonaka, I., & Takeuchi, H. (2000). *Kreowanie wiedzy w organizacji*. Poltex.

Nonaka, I., Toyama, R., Hirata, T., Bigelow, S. J., Hirose, A., & Kohlbacher, F. (2008). Managing flow. In *Managing flow*. https://doi.org/10.1057/9780230583702

Nonaka, I., Toyama, R., & Konno, T. (2000). SECI, Ba and leadership: A unified model of dynamic knowledge creation. *Long Range Planning, 33*.

Paquette, S. (2011). Customer knowledge management. In D. Schwartz & D. Te'eni (Eds.), *Encyclopedia of knowledge management* (2nd ed., pp. 175–184). IGI Global.

Polanyi, M. (1966). The logic of tacit inference. *Philosophy, XLI*(155), 1–18. https://doi.org/10.1017/S0031819100066110

Probst, G., Romhardt, K., & Raub, S. (2000). *Managing knowledge: Building blocks for success*. J. Wiley.

Quintas, P., Lefrere, P., & Jones, G. (1997). Knowledge management: A strategic agenda. *Long Range Planning, 30*(3), 385–391. https://doi.org/10.1016/S0024-6301(97)90252-1

Saviotti, P. P. (1998). On the dynamics of appropriability, of tacit and of codified knowledge. *Research Policy, 26*(7–8), 843–856. https://doi.org/10.1016/S0048-7333(97)00066-8

Schwartz, D. G. (2007). A birds-eye view of knowledge management: Creating a disciplined whole from many interdisciplinary parts. In *Knowledge management in modern organizations* (pp. 18–29). IGI Global.

Skyrme, D. (1997). Knowledge management: Making sense of an oxymoron. *Management Insights, 22*.

Stenmark, D. (2001). Leveraging tacit organisational knowledge. *Journal of Management Information Systems, 17*(3), 9–24. https://doi.org/10.1080/07421222.2000.11045655

Toffler, A. (1980). *The third wave*. Morrow.

uit Beijerse, R. P. (2000). Knowledge management in small and medium-sized companies: Knowledge management for entrepreneurs. *Journal of Knowledge Management, 4*, 162–179. https://doi.org/10.1108/13673270010372297

von Hippel, E. (1999). "Sticky information" and the locus of problem solving: Implications for innovation. *Management Science, 40*(4).

von Hippel, E. (1988). The sources of innovation (Vol. 53, 9). Oxford University Press. doi:https://doi.org/10.1017/CBO9781107415324.004

Wellman, J. (2009). *Organizational learning: How companies and institutions manage and apply knowledge*. Palgrave Macmillan US.

Wijnhoven, F. (2006). Knowledge management: More than a buzzword. In A. Jetter, J. Kraaijenbrink, H.-H. Schroeder, & F. Wijnhoven (Eds.), *Knowledge integration* (pp. 1–16). Springer.

Knowledge Transfer and Knowledge Flows 2

Knowledge in organizations is unevenly distributed. If the knowledge possessed by experts is needed for a given activity of an organization, it must be available at a specific place and time (if it is not available in the required place, it must flow from where it is currently located to where it is needed). The uneven distribution of knowledge justifies the need to organize and improve the flow of knowledge in an organization, ensuring that it is able to fully use its knowledge. In addition, the knowledge accumulated in organizations is often insufficient to run a business and be successful on the market—organizations must also acquire knowledge from their environment (e.g. from customers, partners or competitors) and transfer their knowledge to this environment (e.g. to customers to be able to use the products or services).

Hence, the author examines the subject of knowledge flows, which will be described in this chapter.

2.1 Knowledge Transfer vs. the Concept of Knowledge Flows

Knowledge transfer can be defined as the transfer of knowledge from one place to another, from one person to another, from one organization/department to another, etc. Knowledge transfer may, therefore, consist in, e.g., transferring knowledge from employees to databases, from employees to projects, but also, for example, from CRM (customer relationship management) systems to employees or from organization to organization (e.g. patent sale).

Knowledge transfer is associated with the transmission of knowledge, its assimilation (absorption) and also its use. If knowledge is not used, its transfer is unfounded. Knowledge transfer involves not only replicating the received knowledge by the receiving entity but also using this potentially useful knowledge for its own activities. In the case of transfer, knowledge may be transferred at a given moment (e.g. from an employee to a database), and its use may take place in the

M. Zieba, *Understanding Knowledge-Intensive Business Services*, Knowledge Management and Organizational Learning 10,
https://doi.org/10.1007/978-3-030-75618-5_2

future when it is needed. The transfer of knowledge and its use may, therefore, be separate in time. In the context of knowledge transfer, one should refer to the previously described knowledge conversion processes: socialization, combination, externalization and internalization. In the course of knowledge transfer, knowledge often changes from one type of knowledge to another (e.g. knowledge hidden in an employee's head changes into explicit knowledge by writing down methods of dealing with a client). The employee's knowledge then becomes organizational knowledge and is transferred from the individual to the organization.

Knowledge in an organization flows inside its various elements (structures), between them and between the inside and outside of the organization. According to Sveiby, it is possible to distinguish the following knowledge transfers (Sveiby, 2001):

1. *Between individuals (between employees)*—this type of knowledge transfer is often informal and depends on the willingness of employees to share their knowledge.
2. *From employees to an external structure*—this type of transfer takes place when employees share their knowledge with, e.g., clients. It can put your business at risk if employees share knowledge with competitors.
3. *From an external structure to employees*—this transfer concerns the acquisition of knowledge by employees from external sources, e.g. from customers, suppliers. The knowledge gained in this way can help to improve the company's products/services.
4. *From employees to an internal structure*—this is the transfer of tacit knowledge of employees to the organization, e.g. by submitting ideas for process improvements, coding knowledge in the form of documents and placing them in repositories, etc.
5. *From an internal structure to employees*—an example of this type of transfer are internal training sessions organized by the company for its employees, especially those related to customer service and products/services. This transfer can also take place by familiarizing an employee with the procedures and processes functioning in the organization.
6. *With external structures*—refers to the exchange of knowledge between various types of parties, such as customers, suppliers, distributors, etc., word-of-mouth marketing, where customers informally exchange information about the company and its products.
7. *From an external structure to an internal structure*—an example of this type of transfer is the use of the knowledge of customers and contractors to improve the tools, processes and products of the organization, and even to create innovations.
8. *From an internal to an external structure*—examples of this type of transfer include various forms of advertising and promotion, especially those aimed at educating the client in the area of product/service properties.
9. *Within an internal structure*—this type of transfer is used especially in large, multi-departmental organizations where knowledge rarely flows through informal channels between units (Sveiby, 2001).

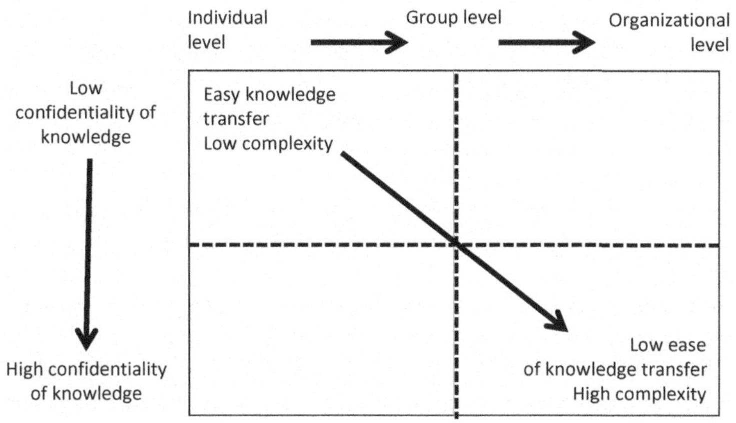

Fig. 2.1 Classification framework for knowledge transfer. Source: Author's own illustration, based on Inkpen and Dinur (1998), p. 457

It is worth paying attention not only to transfers within the internal structure of an organization but also to the exchange of knowledge with the broadly understood environment. As Sveiby notes, nowadays organizations create value based on knowledge, in cooperation with their customers. This new relationship has been called the Value Network and it means the interaction between entities with different roles and connected by different relationships, as a result of which both intangible (e.g. ideas, knowledge) and tangible (e.g. profits) values are created. A natural consequence is a need for effective transfer and conversion of knowledge, which has become the key to building value (Sveiby, 2001).

Knowledge transfer can take place in various ways, be it formal or informal. Knowledge transfer can also include many kinds of knowledge. It can be, for example, knowledge in the form of patents, trademarks or technological solutions, as well as knowledge hidden in the minds of employees. Knowledge transfer can also be based on various types of tools. The choice of the tool depends, among others, on the type of knowledge to be transferred and the organizational level at which the transfer is to be made. The transfer is easier for non-highly classified knowledge and its transfer at the individual level, and most difficult for highly classified knowledge that is to be transferred at the organizational level (Fig. 2.1).

The following are sample tools for transferring explicit knowledge (Kowalczyk & Nogalski, 2007, pp. 106–108):

- Newsletter—tools aimed at notifying users about the appearance of specific knowledge in the system; it is often possible to personalize the received knowledge by selecting topics or areas that the user is interested in.
- Forum, discussion groups, chat—ways to conduct various types of discussions in which all interested parties or a limited number of people with permissions or access keys can participate. Participants can send their comments or opinions on a given topic or in relation to the statements of other users.

- Teleconferences, videoconferences—a concept similar to forums and discussion groups, with the difference that it uses visual and/or sound contact with interlocutors, and thus requires the use of sound or image technology in real time.
- E-mail, instant messaging—solutions enabling the forwarding of messages to one or more users; In addition, instant messaging allows you to obtain information about the status of messages and users.
- E-service or customer service—an automatic customer service system in which customer relationship management (e.g. CRM systems) is of particular importance.
- Document and workflow—solutions based on knowledge management on the plane of information resources exchange between members of an organization. This category includes various types of complex IT systems, often called knowledge management systems by producers, as well as all kinds of databases.
- E-learning training—during this type of training, open, codified knowledge is transferred.

At the same time, there are some sample tools that can be used for the transfer of tacit knowledge (Malamed, 2016):

- Collaboration and social networking—Online collaboration provides a framework for tacit knowledge transfer through a socially constructed learning process. It is based on the assumption that the social nature of the cooperating community gives the opportunity to learn through joint conversations and discourses between participants. Collaborative communities provide opportunities to learn and discover new ideas.
- Showing your work—This is an approach that requires that a specialist's work be visible and shown to others with the intention of transferring the tacit knowledge built into that work. This approach may involve talking about your work as you do it. One example is a surgeon using Google Glass who, with medical students in mind, covers all the steps involved in an operation.
- Storytelling—Organizational stories are considered to be an effective means of capturing and communicating tacit knowledge. Stories transform information into knowledge. They provide the context that gives meaning to facts.
- Recording of lessons learned—Some organizations use an official process of recording lessons learned so that others can take advantage of situations in which they were not involved. It is an effective way of disseminating knowledge gained from experience.
- Mentoring and coaching—these are methods of gaining knowledge from experts, consisting of observation, participation and also performing tasks independently under the supervision of an expert. Thanks to this, a less experienced person can acquire tacit knowledge from a person with extensive experience in a given area and develop their skills.
- Traditional (non-e-learning) internal and external training sessions—during training sessions, tacit knowledge is often transferred to the participants of the session.

Only during face-to-face training can tacit knowledge be transferred. In the case of e-learning training, all knowledge transferred is codified knowledge.

It should be remembered that the basic tool for the exchange of tacit knowledge is interpersonal communication. The mentioned transfer methods are largely based on it.

The selection of knowledge transfer tools and the type of knowledge transferred, as well as the overall approach to knowledge management, should depend on the competitive strategy of an organization: how it creates value for customers, how this value supports the economic model and how employees create value for the organization. *It is worth emphasizing here that knowledge transfer is only one of the elements of knowledge management and further descriptions concern knowledge transfer in the context of knowledge management strategies. Contexting a knowledge management strategy with respect to knowledge transfer is necessary to establish the concept of knowledge flows.* Organizations can use one of two knowledge management strategies: a codification strategy or a personalization strategy (Hansen et al., 1999). Codification strategy involves codifying, storing and re-using knowledge. Knowledge is codified using a "people-to-documents" approach: it is "taken" from the person who developed it, made independent of that person and reused for various purposes. Personalization strategy focuses on a dialogue between individuals, not knowledge objects in the database. Knowledge that has not been codified is passed on through brainstorming and face-to-face conversations. Employees come to solutions together, thoroughly analyzing and discussing the problems they have to solve.

In the case of codification strategies, a characteristic feature is the separation of the person providing knowledge from its recipient: the recipient of knowledge does not have to contact or talk to the supplier directly, but thanks to the codification of knowledge, he can use his knowledge as an independent resource. The codification strategy is appropriate in a situation where knowledge can be reduced to an overt form. Companies adopting this strategy will invest heavily in IT solutions that can collect large amounts of explicit knowledge and make it available to employees for reuse. In summary, the codification strategy is based on the reuse of codified knowledge.

In the case of a personalization strategy, contact (personal, by telephone, online, etc.) between the sender and recipient of knowledge is necessary. Without this element, it is impossible to transmit tacit knowledge between the sender and the recipient. This strategy is based on employees' individual expertise which cannot be codified and put into a database. Thanks to this knowledge, employees are able to come up with new, creative solutions that are unique and cannot be universally duplicated, as in the case of codification strategies. The necessity of direct interaction between people (members of an organization) results in the development of networks connecting them, enabling conversations and the exchange of tacit knowledge.

When analyzing the strategies of codification and personalization, it can be seen that in the case of codification strategies we deal (most often) mainly with the transfer of explicit knowledge, and in the case of personalization strategies—with the transfer of tacit knowledge, although this cannot be regarded as a hard and fast principle. In the personalization strategy, explicit knowledge can be transferred, and tacit knowledge can be transferred in the codification strategy, although these will not be the dominant types of knowledge. Different knowledge transfer tools will be used for each type of knowledge management strategy. Moreover, when transferring tacit knowledge, it is possible to transfer it by third parties (e.g. one employee may transfer tangible knowledge from another employee to a third employee), and when transferring tacit knowledge, it is not possible.

The approaches to KM discussed in the previous part of the work (e.g. the process approach) are used by an organization to obtain as much knowledge as possible, to codify it and make it available to people who need it. In order for an organization to transfer knowledge on a large scale, it must be codified and placed on some type of medium (e.g. in a database). Then it can be made available and easily transferred to employees who need it.

However, the current pace of the growth of knowledge and its continuous development is so fast that such an approach is in many cases difficult or even impossible to implement. The reasons for this may be various, such as time constraints (codification of knowledge, especially of tacit knowledge, is very time-consuming). It also happens that the amount of work and financial resources is disproportionate to the benefits obtained, and the stored knowledge quickly becomes outdated. Not every organization can afford to implement a complex knowledge management system with the help of which it will acquire and store various types of knowledge possessed by employees. This kind of approach to knowledge management is justified in some organizations, but cannot be implemented in all of them.

Overall, in today's organizations, the following challenges related to knowledge and management can be identified:

1. Knowledge is very often needed here and now; there is no time to search, code and share it—it must be free continuous access. This is well illustrated by the concept of knowledge "on demand", in which knowledge must be available whenever it is needed. In other words, access to knowledge should be continuous so that it can be used whenever it is needed.
2. Knowledge is born and develops in human minds (Davenport & Prusak, 1998, p. 5). Moreover, it is knowledge that enables taking actions and decisions, both at the level of an individual and at the level of an organization. For action and decision-making to be possible, knowledge must not be missing, it must be made available wherever it is needed.
3. In every organization, there are large amounts of explicit and tacit knowledge, located in different places. Tacit knowledge often cannot be easily codified and made available at the organizational level. Therefore, it is necessary to create appropriate conditions in an organization to ensure that this knowledge can flow freely and be available wherever it is needed.
4. KM solutions or KM systems, which are characterized by a high level of structuring and formalization, may not always be introduced, especially when it

comes to small- and medium-sized enterprises (SMEs). Research shows that small companies often introduce KM in an unplanned manner, testing available solutions and quickly adapting them to their needs (Zieba et al., 2016).
5. Constant development and the growing availability of IT tools make more and more companies invest in this type of solutions, sometimes forgetting the basic method of transferring knowledge (through individual conversations). The key issue is the willingness to share knowledge (one way or another), and IT tools only serve as a medium for this kind of activity.
6. Even the best solutions will not be used if employees do not want to do so. In addition, IT tools are widely available to all companies, and it is more difficult to build an organizational culture that fosters knowledge sharing and dissemination.

Taking into account the above challenges, it is proposed to introduce the concept of knowledge flows:

The flow of knowledge is a kind of voluntary flow,[1] *knowledge transfer involving two parties: the sender and the recipient of knowledge. In other words, the flow of knowledge can be defined as the movement of knowledge between people.*

For the flow of knowledge to take place, two elements are necessary: people who pass on their knowledge and those who receive it and a difference in levels of knowledge (the knowledgeable person knows something that the receiving person does not know). If any of these elements are not present, knowledge cannot flow. Knowledge flows occur only if both the recipient of the knowledge and its sender want it. Importantly, the transfer of knowledge from the sender to the recipient does not result in the disappearance of knowledge in the sender, but a kind of duplication/multiplication of this knowledge, so that it is found both in the sender and in the recipient. The flow of knowledge is what happens between the sender and the recipient of knowledge—there is a thread of understanding between them—one transmits his knowledge in some form, and the other receives and absorbs it, combining it with the knowledge already possessed and thus creating new knowledge. It should be noted that the flow of knowledge may be two-way, i.e. the roles of the receiver and the knowledge sender may be parallel. Then the knowledge sender will have the knowledge that interests the recipient who does not have it, and vice versa (Fig. 2.2).

Knowledge flows allow responding to the above-mentioned challenges because they enable free, continuous access to the knowledge of colleagues, clients, partners and other parties. *Thanks to the flow of knowledge, knowledge can be made available wherever it is needed, when it is needed.*

The flow of knowledge in an organization can also be defined in a very broad sense—it then means all kinds of knowledge-related changes taking place in the

[1]The voluntary transfer of knowledge is important in the sense that if a participant does not wish to transfer or receive knowledge, the transfer cannot take place. For example, if employee X gives employee Y a document that employee Y does not want to read, the transfer will fail. Similarly, if worker X needs knowledge from worker Y and the latter does not want to pass it on to him, there will be no knowledge transfer either.

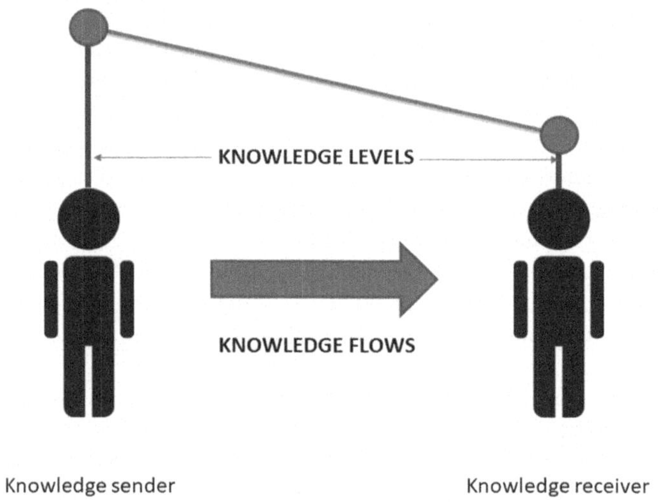

Fig. 2.2 Knowledge flow. Source: Author's own illustration

organization, such as knowledge conversion, knowledge transfer, knowledge sharing, knowledge integration, knowledge reuse and other processes relating to changes, movements and applications of knowledge over time. In this approach, the flow of knowledge was presented by Sveiby (2001), who distinguished the nine streams of knowledge flows described in the previous section. The introduced concept of knowledge flows does not oppose this approach but complements it. The overall flow of knowledge in an organization is the result of all activities related to knowledge, its transfer, transformation, integration, use, etc., including knowledge transfers (both explicit and hidden). In other words, the overall flow of knowledge also includes individual flows of knowledge that take place between members of the organization and representatives of its broadly understood environment.

The phenomenon of knowledge flows is rather little discussed. Additionally, the few publications on this subject are often of practical nature (e.g. Leistner, 2011; Nissen, 2006) or they are focused on solutions in the area of information technology (Trentin, 2011). At the same time, all publications in this area to date focused on internal knowledge flows in organizations (e.g. Nissen, 2005; Zhuge, 2002), ignoring the aspect of knowledge exchange with the broadly understood external environment (e.g. customers, suppliers, distributors, R&D units or universities). Later in this subsection, there will be presented the links between the concept of knowledge flows and other terms used, as well as the characteristics and taxonomy of knowledge flows.

2.2 The Concept of Knowledge Flows Versus Other Concepts

Knowledge flow vs. knowledge sharing. One of the elements of knowledge flows is knowledge sharing. Knowledge sharing is the voluntary sharing of knowledge by an employee with other employees or organizations. Knowledge sharing can take place in one or both directions. However, in knowledge sharing there is no need to use knowledge—this is one of the differences between knowledge flow and knowledge sharing. For example, employee X may pass on (share) his knowledge of how to encourage a customer to buy with other employees, and that knowledge may or may not be used by them. In this case, the fact of sharing/exchanging knowledge is important, and not its acquisition, application or implementation in an organization. In the case of the flow of knowledge, an important element is the willingness to share knowledge on the part of the person possessing it (willingness to share knowledge) and the need to receive knowledge from the recipient.

The flow of knowledge and the classic perception of resources and flows in an organization. The distinction between stock and flows comes from economics. The resources and streams of knowledge can be illustrated with the "bathtub" metaphor: at any time we can check the water level in the bathtub, which is the cumulative result of water inlet (through the tap) and outflow (through the drain). For example, in the case of research and development, the amount of water in a bathtub represents the pool of expertise at a given point in time, while the knowledge gained from ongoing R&D expenditure is illustrated by the tap water flowing in, and the fact that expertise ages over time. is illustrated by the drainage of water through the drain (Dierickx & Cool, 1989). Knowledge resources are accumulated knowledge assets that are inside the company, and knowledge streams flow into the company or are located in different parts of the company and can be assimilated and developed into knowledge resources (DeCarolis & Deeds, 1999).

Streams of knowledge (its inflow and outflow) are aimed at increasing/reducing knowledge resources in an organization. In turn, knowledge resources define the accumulated explicit and implicit knowledge available in the organization. A resource of knowledge may be in an open form, such as a repository or organizational library, or hidden, such as the knowledge of employees, their customer service skills, etc. Knowledge resources may be distributed throughout the organization, in various places and forms. The stock of knowledge is a static approach to knowledge in the sense that it informs about the knowledge gathered and possessed here and now by a given organization.

In this approach, the stream of knowledge is understood broadly and covers the movement of knowledge both within an organization and between the organization and the environment. The flows of knowledge will be both its inflows and outflows. If an organization loses an employee, e.g. as a result of his retirement, it will be a stream of knowledge (or specifically its outflow) that reduces the knowledge resources in the organization. If the organization purchases a patent, it will also be a knowledge stream (inflow), which in turn increases the organization's knowledge resources.

In this work, the flow of knowledge has a different meaning than the stream of knowledge and is subject to limitations (e.g. it is related to the absorption of knowledge by the recipient). In this sense, the flow of knowledge serves to share knowledge (e.g. possessed by an expert in an organization).

Knowledge flow vs. workflow. Knowledge flows are often linked to the workflow of an organization. A workflow can be defined as the progress of the various stages (tasks, events, interactions) that make up the work process, involve two or more people, and create or add value to the organization's activities.[2]

In some situations, it is necessary to generate a knowledge flow before starting work or performing a specific task. If this does not happen, an employee of an organization will not be able to perform his tasks. Sometimes the flows of knowledge have to run parallel to the implementation of tasks—e.g. if the project is a multi-stage project, and to start the next stage, it is necessary to complete the previous stage and transfer knowledge about the completed tasks. The flows of knowledge that must take place before the implementation of specific tasks can be described as sequential (first, the flow of knowledge takes place, and then the task is completed). On the other hand, when the flows of knowledge take place at the same time as the execution of tasks, they can be described as parallel. Regardless of the time horizon, knowledge flows are related to the workflow also in the sense that workflows enable new knowledge flows—after completing a task, an employee or team gains new knowledge that can be transferred to other employees/units and used to perform subsequent tasks. Overall, streamlining knowledge flows and linking them better to the workflow can contribute to better performance by an organization. Sequential and parallel flows of knowledge are presented in Fig. 2.3.

2.3 The Subject of Knowledge Flow and Its Stages

An important aspect of the flow of knowledge is the type of knowledge it relates to. On the one hand, in order to build a knowledge-based competitive advantage, it is necessary to define essential knowledge, which is a strategic asset of the enterprise. In other words, not all knowledge is equally important and requires equal attention. On the other hand, not only key knowledge or knowledge that is the source of competitive advantage must flow freely in an organization—the knowledge needed to perform ordinary, everyday tasks by employees is also important in this context. Lack of this knowledge or difficulties in its flow may result in the loss of competitive advantage (Nissen, 2006, pp. 9–10). Referring to the already mentioned concept of competitive advantage, it can be stated that tacit knowledge is usually a better source of competitive advantage than explicit knowledge (von Hippel, 1999).

In the flow of knowledge, the form of knowledge may remain the same, and it may also change. For example, if knowledge flows from one employee to another, it

[2]Business Dictionary, http://www.businessdictionary.com/definition/workflow.html (date of access: 06.11.2017).

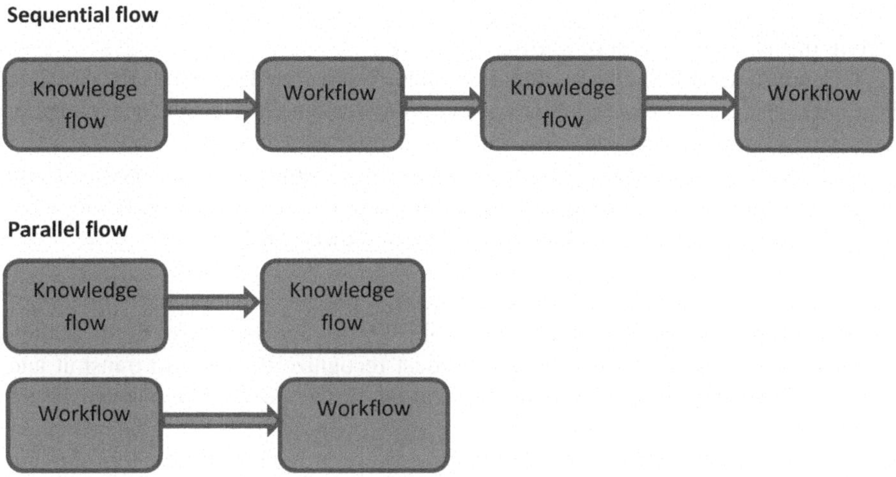

Fig. 2.3 Sequential vs. parallel flows of knowledge. Source: Author's own illustration

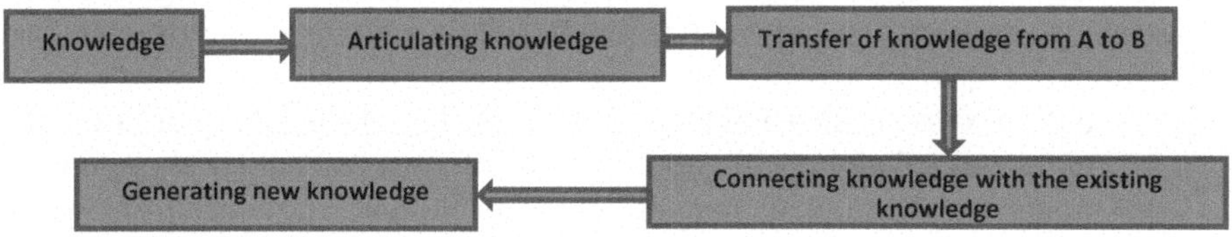

Fig. 2.4 Knowledge in the flow of knowledge. Source: Author's own illustration

can transform from tacit knowledge of an employee into explicit knowledge, transferable to another employee, and after flowing to another employee, connect with his existing knowledge and resume the form of tacit knowledge. This illustrates the already mentioned concept of dynamic knowledge, which not only flows between people, places, in time, etc., but also changes its form, develops and integrates with the encountered environment. We can use here an analogy with a river which, flowing from place A to place B, changes its form (e.g. width or depth), and also "takes" encountered elements (e.g. branches) with it, which change its form. It is similar to knowledge which, flowing inside and outside an organization, undergoes various processes, changing its form (from hidden to explicit and vice versa), connecting with other knowledge, or becoming the basis for generating new knowledge (Fig. 2.4).

The flow of knowledge is an element that integrates the above phenomena in the sense that knowledge would not be articulated if it were not to flow, and new knowledge would not have been generated if it had not been for the flow of existing knowledge and its connection with the knowledge available in a new place. In the analysis of knowledge flows, it is not enough to focus only on the technical aspect of changing the location of knowledge from point A to point B, but also the phenomena preceding and following it. An analysis of knowledge flows in the context of the technical transfer of knowledge from point A to point B would offer an incomplete picture of the phenomenon of knowledge flows.

To analyse knowledge flows, it is important to separate their stages. Knowledge flows are inherent in broadly understood communication, and the basic elements of any type of communication between two people are message, sender, coding scheme, channel, transmission over the channel, decoding scheme, recipient and assigning meaning to a decoded message (Gupta & Govindarajan, 2000).

In this work, ten phases of knowledge flow are distinguished.

Stage 1. In the first stage, there must be a need to receive knowledge on the part of the recipient and the willingness to send this knowledge on the part of the sender. At this stage, the sender and receiver must recognize the need to transmit and receive the knowledge. Without this condition, the flow of knowledge is not possible.

Stage 2. The second stage is the preparation of knowledge for the transfer. In many cases, it will be necessary to simplify knowledge or partially codify it in order to enable its efficient flow. For example, if employee A is tasked with passing on his knowledge of customer service to employee B, he will have to organize this knowledge properly and at least partially codify it so that employee B can acquire it.

Stage 3. At the next stage, the knowledge flow channel is selected. In most cases, the most effective solution will be the use of non-technological solutions (e.g. organizing traditional training or a mentoring relationship between employee A and employee B) or, possibly, partial support of the traditional relationship between employees with some technological tool.

Stage 4. The next step is to code or articulate knowledge in such a way that it will be passed on to the recipient. This stage follows the selection of the knowledge flow channel because the channel to some extent determines the method of coding knowledge (e.g. in a mentoring relationship, knowledge is passed on orally or in the form of practices).

Stage 5. Only after selecting the channel and coding the knowledge in an appropriate way is it possible for the sender to send the knowledge.

Stage 6. At this stage, the recipient receives the knowledge.

Stage 7. After receiving the knowledge, the recipient decodes the knowledge.

Stage 8. At the next stage, he or she assimilates it. Sometimes there are disturbances in the transmission of knowledge and the recipient is unable to understand or decode it. Then he or she has to ask the sender to re-send the knowledge or the missing parts.

Stage 9. The next step in the flow of knowledge is to provide feedback on the knowledge and its flow. This step is sometimes skipped, although it can help improve knowledge flows. If there were any problems with the transfer of knowledge, they can be eliminated or minimized in the future thanks to the feedback.

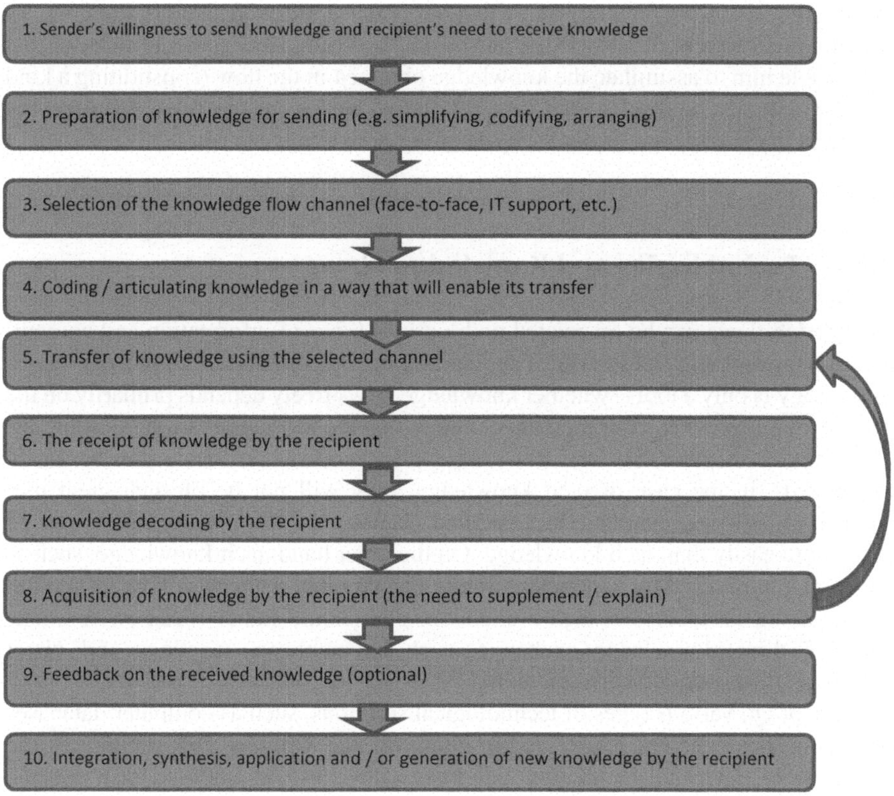

Fig. 2.5 Knowledge flow stages. Source: Author's own illustration

Stage 10. In the last stage, the recipient integrates, synthesizes or applies the obtained knowledge. Knowledge flow can also generate new knowledge based on the knowledge received by the recipient.

These stages are presented in Fig. 2.5.

From the point of view of an organization, the last stage is extremely important in the flow of knowledge, because it allows you to benefit from the transferred knowledge. Knowledge in an organization is a potential source of competitive advantage, and if used well, it contributes to achieving better results. Therefore, the possibility of a broadly understood use of the acquired knowledge (through integration, synthesis, application and generation of new knowledge) is the meaning of knowledge flow. Usually, the recipient already has some knowledge that he needs for the receipt of the knowledge that flows to him. At the same time, the flowing knowledge is missing knowledge, previously not possessed by the recipient. Without preparation (having appropriate knowledge), the recipient would not be able to "pick up" and absorb this knowledge. Moreover, the flows of knowledge are inherently related to learning. The recipient of the flowing knowledge increases his knowledge, acquiring the knowledge that has reached him as a result of the flow. To sum up, the recipient of knowledge must usually already have some knowledge that will enable him to assimilate the knowledge obtained in the flow (constituting a kind of "foundation"), and after receiving the knowledge from the flow, the recipient increases his knowledge resources by learning new things.

2.4 Technologies and Knowledge Flows

Knowledge flows can be associated with various types of information and communication technologies. However, it should be remembered that in knowledge flows, technology is only a tool—whether knowledge flows freely depends primarily on the people (employees) in an organization, and not on the technology used. IT solutions can support, in particular, the flows of explicit knowledge which can be easily articulated—in the case of tacit knowledge, they will not be of such great use. Explicit knowledge, available in a codified version, is transferred using various IT tools more easily than tacit knowledge. On the other hand, tacit knowledge, such as knowledge possessed by a member of an organization or a client, will be difficult to extract using IT tools—in such a situation, one should rather focus on solutions enabling direct knowledge exchange, such as employee meetings and direct exchange of knowledge between them.

Very often, various types of technological solutions, such as computer databases or repositories, allow for the organization, storage, transfer and support of the exchange of data and documents, but these data must be analysed in a specific context so that they can be turned into knowledge. It can be concluded that these solutions constitute a specific infrastructure of an organization, helping it to manage data, information and partly knowledge (its explicit part). In order to be able to process, transmit, display and otherwise use knowledge with IT tools, most of the content must be clearly articulated (e.g. as a written text, in graphical form, as a mathematical formula or software code). E-mail, text chat, Internet search, social networking and all kinds of mobile applications are becoming more and more common. Related knowledge (e.g. text or e-mail content, the material found in search engines, tweets or website publications) is generally public knowledge. In turn, evolving opportunities to facilitate remote and multilingual collaboration in AV and shared applications increase the potential to support tacit knowledge flows as well (Nissen & Bordetsky, 2011, p. 3).

To sum up, it can be stated that in the case of tacit knowledge transfer, the right choice will be various types of non-technological solutions, such as meetings, mentoring or coaching, because tacit knowledge is best shared in direct interpersonal relationships, and not with the use of IT tools. Tools (e.g. communicators) which are a channel of communication and enable direct exchange of knowledge, regardless of the location of people participating in this exchange may be an exception. In turn, for the needs of explicit knowledge flows, it is worth using IT solutions.

2.5 The Flow of Knowledge and the Flow of Data and Information

Knowledge flows should also be discussed in relation to data and information flows (Fig. 2.6). First of all, it should be noted that knowledge differs from information and data in terms of enabling organizations to gain a competitive advantage. The transfer of information and data via computers, networks, reports and messages does not create a flow of knowledge, at least not directly (Nissen, 2006, p. 12). According to Nissen (2005), knowledge flows are closely related to information and data flows: data flows support information flows, and these in turn partially determine knowledge flows. At the same time, knowledge flows do not translate directly into the results of an organization (they are difficult to measure and noticeable), but by supporting workflows they contribute to the improvement of the achieved results.

This work focuses mainly on knowledge flows, as data and information and their flows are relatively easy to identify and improve. This is due to the fact that data and information are usually easily transferable and more accessible than knowledge (especially hidden). At the same time, it is the hidden knowledge that is often the source of competitive advantage due to its limited transferability—the competition will find it easier to obtain data or information from the company than to take over its hidden organizational knowledge. Therefore, this monograph analyses the flows of knowledge that may take place not only inside but also outside the company (flows to and from customers, partners, suppliers, R&D centres, etc.).

The concept of knowledge flows is, according to the author, useful for explaining phenomena related to knowledge occurring in an organization and not covered by the analysis within the traditional approach to knowledge management. It should be

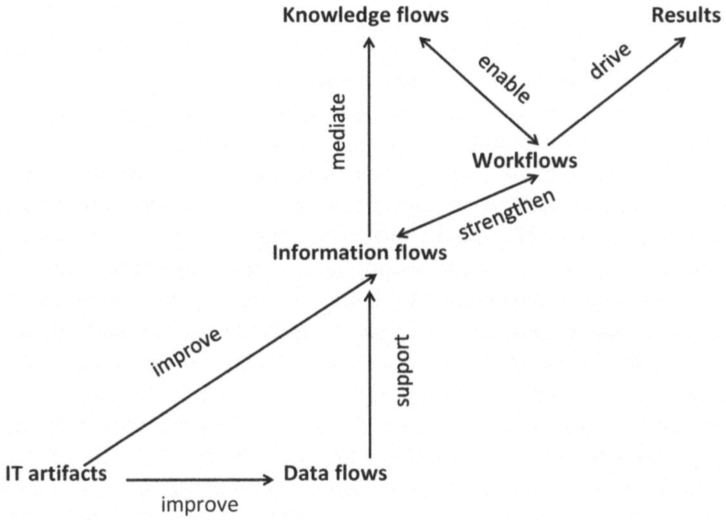

Fig. 2.6 The relationship between knowledge flows, data and information flows, and workflows. Source: Author's own illustration, based on: Nissen (2005), p. 255

noted that the concept of knowledge flows is not in opposition to the knowledge management trend, but it is rather an attempt to supplement it. As Prusak (2003) stated, knowledge flows through organizations along existing paths. If we want to understand how to improve this flow, we need to understand these paths. Knowledge flows underline the importance of dynamic knowledge that often changes its form. They also explain how knowledge moves inside an organization and between an organization and its environment. Knowledge flows also contribute to the accumulation, exchange, updating, combination and replenishment of knowledge. Being aware of knowledge flows and how to improve them can help organizations manage knowledge more effectively and gain and maintain a competitive advantage.

2.6 Knowledge Flow Taxonomy

The taxonomy comes from the combination of the Greek words taxis, or order, and nomos, or law, norm, rule (Currás, 2010, p. 37). The term "taxonomy" means generally "the principles or conventions of order or arrangement"(Lambe, 2007, p. 4). There are several different forms of representation: trees, hierarchies, matrices, facets and system maps. The most basic form of taxonomy are lists; listing is also a basic task in most other (more complex) ways of representing taxonomies. Lists classify related items into categories and show the similarities between them. The tree structure is a technique for visualizing and organizing relations between taxonomies (Lambe, 2007, pp. 14–18).

Hierarchies, in turn, are a special type of tree structure. They can be represented as pyramidal structures, they can have certain levels, but they do not have to. Matrices work best in well-defined areas of knowledge that can be organized along two or three dimensions. They can help you understand categories and highlight gaps or missing categories. Aspect (faceted) taxonomies built on aspects hierarchize a set of taxa, assuming their various characteristics as the criteria for their classification. Typical aspects can be people, organizations, themes, products, locations or activities. System maps are visual representations of a field of knowledge in which closeness and connections between categories, as well as relationships between real phenomena are expressed (Pellini & Jones, 2011, pp. 10–13).

Two types of system maps can be distinguished: descriptive maps and conceptual maps. Descriptive maps are excellent tools for transferring knowledge about a given domain in its natural context. An example of this type of map is the taxonomy of arteries and veins in humans, shown in the form of a map. Conceptual maps, on the other hand, are about mental, not physical, constructs. The most common form of a conceptual map is a mind map in which different ideas are linked by branched lines. Proximity and connections are easier to represent on such a map than in a tree structure. Mind maps also often include colour coding to indicate discrepancies between different categories. Concept maps are a more rigorous form of the mind map. In the concept map, the nature of the relationship between the two concepts is clearly defined. Concept maps are extremely useful in cataloguing specialized knowledge domains as well as explaining specific domains of knowledge in the form of a beginner's summary (Lambe, 2007, pp. 42–44).

There are different types of knowledge flows. According to the author, their division can be made according to several dimensions, such as flow direction, flow duration, time constraints, flow location, relationship between the sender and recipient of knowledge, flow level, flow planning, flow formalism and the initiating party (push/pull approach), form of knowledge, area of knowledge, the relevance of knowledge. Below is a description of individual dimensions along with the categories of flows distinguished on their basis.

1. *Direction of flow*—this dimension determines in which direction(s) the knowledge flows. Based on the direction, two types of flows can be distinguished:
   a. one-way flow that occurs between two entities (e.g. a company employee and a customer) in only one direction (e.g. employee to customer or customer to employee);
   b. two-way flow that takes place between two entities (e.g. a company employee and a client) in both directions (e.g. between a client and a company employee, and between a company employee and a client).
2. *The duration of the flow*—this dimension determines whether the flow takes place at a given moment in time or is extended in time and the transfer of knowledge takes place at a different time than its reception by the recipient. Within this dimension, two types of flows can be distinguished:
   a. synchronous flow—a flow in which there is the unity of time and place of knowledge flow , e.g. face-to-face conversation between employees;
   b. asynchronous flow—a flow in which there is no unity of time and place of knowledge flow, i.e. the message of knowledge does not coincide with its reception in time, e.g. a company employee watches a video in which a colleague presents good practice in an organization.
3. *Flow time limits*—this dimension determines whether the flow of knowledge must take place within a certain time (it is urgent) or there is no time limit, and participants in the flow of knowledge can use as much time as needed for the flow. Two types of flows can be distinguished here:
   a. urgent flow—there are time limitations in the flow of knowledge, e.g. late execution of the client's order and the need to quickly settle the matter using the knowledge obtained;
   b. normal flow—no time limits in the flow of knowledge.
4. *Location of the flow*—this dimension indicates where the flow is taking place—whether within or outside the boundaries of an organization. Within this dimension, two types of flows can be distinguished:
   a. internal flows within an organization (e.g. within a department, between departments, between individual employees);
   b. external flows between an organization and its environment (e.g. customers, contractors, partners, etc.).
5. *The relationship between the sender and recipient of knowledge in the flow*—in this dimension, it is possible to define the type of relationship between the unit

transmitting knowledge within the flow and the unit receiving knowledge. The types of this relationship are as follows:

a. mentoring relationship in the flow of knowledge (e.g. when an older employee transfers his knowledge to a younger one who has no experience in a given area yet);
b. coaching relationship in the flow of knowledge (e.g. when an employee works with a coach, trying to develop his professional abilities);
c. peer relationship in the flow of knowledge (e.g. when two employees exchange knowledge in an informal way and not related to their work relationship);
d. management relationship in the flow of knowledge (e.g. when the supervisor requires the employee to disclose knowledge regarding customer observation).

6. *The level of flow in an organization*—this dimension determines the level of knowledge flow in an organization. The following levels can be distinguished:
   a. individual flow of knowledge—occurs between two people, e.g. between two employees or between an employee and a client;
   b. group/team knowledge flow—takes place between a group/team of people (e.g. employees) and an individual (e.g. during a meeting between the superior and the employees);
   c. intergroup/inter-team knowledge flow—takes place between one group (team) of people (e.g. employees) and another group (e.g. employees), e.g. when one group (team) cooperates with another and they have to share knowledge on the progress of work.
7. *Flow planning*—it determines to what extent the flow is a planned activity and to what extent it is spontaneous. Within this dimension, the following can be distinguished:
   a. intentional/planned flows—flows that are planned, placed in the structure of an organization and take place regularly (e.g. employee meetings);
   b. incidental/ad hoc flows—flows that are not planned and take place spontaneously (e.g. employee coffee meeting).
8. *Flow formalism*—this dimension shows to what extent the flow is a formalized and structured activity. This dimension is linked to the previous dimension (flow planning) and specifies:
   a. formal flows (e.g. mentoring system for newly hired employees);
   b. informal flows (e.g. staff meeting after hours).
9. *Party initiating flows (push/pull approach)*—this dimension determines who is the initiator of the flow of knowledge—whether it is its owner (knowledge sender) or a person who needs it (knowledge recipient). Within this dimension, the following can be distinguished:
   a. push flow—initiated by the knowledge holder;
   b. pull flow—initiated by a person who needs knowledge.
10. *Form of knowledge*—within this dimension, it is determined what form of knowledge is passed on in the flow—whether it is tacit knowledge, overt knowledge or a combination of these two types of knowledge:

   a. tacit knowledge flow—the flow of tacit knowledge;
   b. the flow of explicit knowledge—the flow with explicit knowledge as its subject matter;
   c. the flow of explicit and tacit knowledge—the flow with explicit knowledge and tacit knowledge as its subject matter.
11. *Knowledge area*—this dimension determines from which area comes the knowledge transmitted in the flow. Within this dimension, flows can be divided into:
   a. flows of general knowledge (e.g. on new technologies, legal situation, etc.);
   b. business knowledge flows (concerning the market, customers, organization and its broadly understood functioning).
12. *The importance of knowledge*—this dimension defines how important the flow of knowledge is for the functioning of an organization. In this dimension, the following flows can be distinguished:
   a. key knowledge flows;
   b. flows of useful, but not essential knowledge;
   c. flows of irrelevant knowledge.

Based on the presented dimensions of knowledge flows, a taxonomy of these flows was developed. The taxonomy of knowledge flows in an organization will be presented in the form of a conceptual map, because it enables the organization of individual categories of flows (Fig. 2.7).

2.7 Factors Influencing the Flow of Knowledge

Many factors can influence the flows of knowledge. Free flow of knowledge will be difficult in the case of the so-called internal stickiness of knowledge, which is a metaphor for the difficulties encountered in transferring knowledge within an organization. Szulanski (1996) identified a number of reasons influencing the intrinsic stickiness of knowledge and grouped them into four categories: the characteristics of the transferred knowledge, the characteristics of the source of knowledge, the characteristics of the recipient of knowledge and the characteristics of the context.

The characteristics of the transferred knowledge include the ambiguity of knowledge (it can be received/understood in various ways in an organization and it can even adversely affect its activities) and the lack of verification/confirmation of knowledge (no proof of the validity/usefulness of the knowledge to be transferred). The characteristics of knowledge are also related to its form—whether it is tacit (more difficult to convey) or overt (easier to pass) knowledge.

The characteristics of knowledge sources include the lack of motivation and the lack of perceiving the person as a reliable source of knowledge. Lack of motivation means that the source of knowledge may be reluctant to share relevant knowledge for fear of losing property, privileges and superiority; you may also find that the individual is not properly rewarded for sharing hard-won success or is reluctant to devote time and resources to help the flow of knowledge. The problem may also be

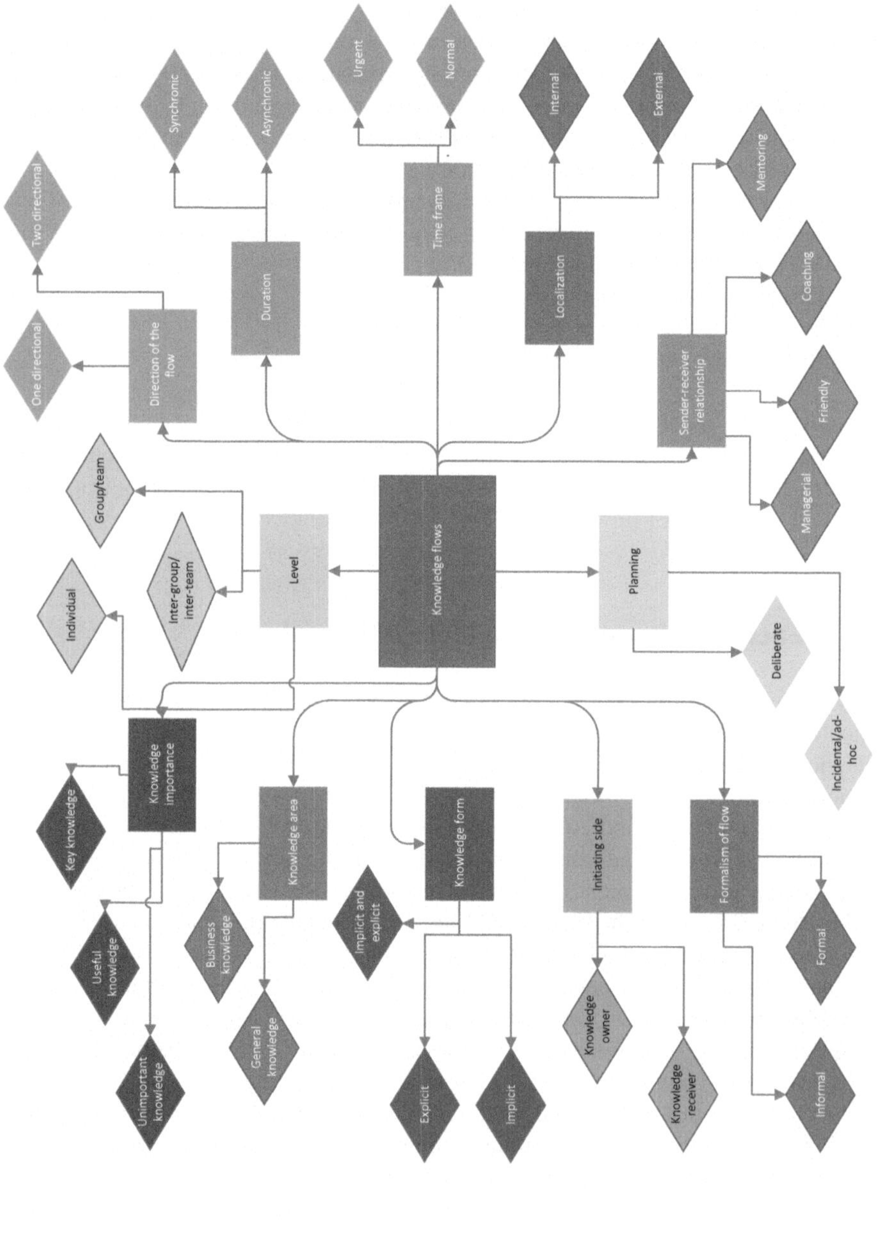

Fig. 2.7 Conceptual map of knowledge flows. Source: Author's own illustration

the lack of perception of a person as a reliable source of knowledge—if the source of knowledge is not seen as reliable, is not considered trustworthy or competent, initiating a flow from that source (from that person) will be more difficult, and its advice and examples may be questioned.

The characteristics of the recipient of knowledge are related to three aspects: lack of motivation, lack of absorptive capacity and lack of retention capacity. Lack of motivation of the recipient of knowledge may result in passivity, mock acceptance, hidden sabotage or complete rejection in the implementation and use of new knowledge. The recipient's lack of absorptive capacity means that he or she cannot accept and use the received knowledge, and the lack of retention capacity is related to the inability to retain knowledge and integrate it with the existing knowledge. The issue of the recipient's absorptive capacity and retention capacity is very important in knowledge flows. In a situation where the recipient is not able to absorb and integrate the knowledge that has been transferred to him in the framework of the knowledge flow, he will have to bear the costs of this transfer without being able to use the knowledge obtained. If this is taken into account, only those flows where the recipient is able to absorb and use the knowledge under consideration are justified.

The characteristics of the organizational context mean its sterility and burdensome relationships. The sterile organizational context makes the flow of knowledge difficult because knowledge falls on fertile ground. Formal structure and systems, sources of coordination and expertise, and attributes related to the organizational context influence the number of transfer attempts and their results. Burdensome relationships additionally hinder the flow of tacit knowledge, which often requires communication and direct exchange (Szulanski, 1996).

Gupta and Govindarajan (2000) identified similar factors at the level of organizational units, not of individual employees. According to these researchers, the following factors influence the flows of intra-organizational knowledge between units of a given organization (to or from a subsidiary/branch of international corporations):[3]

- value of the source unit's knowledge resources;
- the source unit's motivational tendency;
- existence and variety of transmission channels;
- motivational tendency of the target entity;
- absorptive capacity of the target entity.

As for the first factor, the value of knowledge resources, it is well known that not all knowledge is equally valuable to an organization (Barney, 1991). Therefore, the greater the value of the knowledge possessed by a given unit (team/branch), the

[3]Although these researchers analysed the transfer of knowledge between subsidiaries of multinational corporations, the factors they identified may also be applied to the analysis of knowledge flows within other organizations, as well as between organizations and their environment.

greater its attractiveness and willingness to be acquired by other units, and, consequently, the greater the pressure on the knowledge-equipped unit.

The second factor, the source unit's motivational tendency (analogous to the one described earlier), concerns the individual's willingness to share their knowledge and to organize the flow of knowledge. Having unique knowledge often gives a favourable competitive position within an organization, and therefore the individuals concerned may intentionally hide their knowledge and not want to transfer it as part of the knowledge flow to other places in the organization. Motivating factors and appropriate incentives for individuals to actively participate in knowledge flows play an important role at this stage, potentially offsetting the losses related to the transfer of their knowledge.

The third factor, the existence and variety of transmission channels, obviously influences the flows of knowledge in the sense that knowledge cannot flow if there are no channels that enable it. Apart from the mere availability of transmission channels, their properties are also important, such as the richness and bandwidth of the communication, expressed in informality, openness and channel density.

Another, fourth factor, the motivational tendency of the target entity (also mentioned earlier at the individual level), is related to the readiness and willingness of this individual to receive knowledge from the flow. The main problem in this area is the "Not-Invented-Here Syndrome" (NIH Syndrome), which is the result of the mechanisms of self-defence by managers and the struggle for power in organizations. In short, some individuals may not be willing to accept the knowledge transferred because they prefer to generate their own knowledge and show that they are able to do so.

The last, fifth factor, the absorptive capacity of the target entity, is related to the aforementioned ability to acquire knowledge by the recipient (in this case, the team/branch of an organization). The absorptive capacity to acquire knowledge is the ability to recognize the value of new knowledge, assimilate it and use it for commercial purposes (Cohen & Levinthal, 1990). This ability depends on two factors: the previously possessed knowledge base, allowing for the assimilation of the acquired knowledge, and the so-called homophilia[4] (similarities) between the unit transmitting the knowledge and the unit receiving it. Previously acquired knowledge is important because it shapes the filters by which an organization distinguishes between more important and less important knowledge and because it determines an organization's ability to internalize and absorb more valuable knowledge (Cohen & Levinthal, 1990).

The first two factors (the value of the source unit's knowledge resources and the source unit's motivational tendency) determine the existence of the knowledge flow from a given unit, and the last two (the target unit's motivational tendency and the

[4]Homophilia—the degree to which two or more people who interact with each other is similar in some characteristics, such as beliefs, education, social status, etc. The greater the similarity between two individuals, the greater the potential knowledge gain, more effective shaping of the approach and more visible change of behaviour (Rogers, 1995, pp. 18–19).

target unit's absorptive capacity) influence the reception of these flows by the individual—the recipient. The middle factor, the existence and variety of transmission channels, affects both (Gupta & Govindarajan, 2000).

Apart from those identified above, there are other factors influencing the flow of knowledge. Knowledge flows are based on knowledge sharing, because the knowledge sender makes his knowledge available to other units, and thus shares his knowledge. Therefore, the flow of knowledge is influenced by factors determining the sharing of knowledge (apart from the above-mentioned motivational tendency of the source unit), which include: organizational culture, trust, management support and the competences of leaders[5] (McDermott & O'Dell, 2001; Tseng et al., 2012).

Organizational culture is a very important aspect that supports or hinders the sharing of knowledge. If employees are aware that sharing knowledge is an important element of their work, which is reflected in the incentive system and employee appreciation, they will be more willing to participate in it. In a culture where the value of knowledge is appreciated, an organization is supported in learning by many factors, such as availability of knowledge, knowledge exchange, knowledge flow, IT infrastructure, personal networking, systems thinking, leadership, communication climate, problem solving and training (Tseng et al., 2012). Moreover, a collaborative organizational culture supports the exchange of knowledge between its members (Goh, 2002).

Trust, and especially interpersonal trust as a key aspect of relational capital, is important as it can strengthen the density of the organizational network, thus leading to a higher level of knowledge sharing within an organization (Hsu & Chang, 2014), and therefore an easier flow of knowledge. Trust is necessary to persuade people to share their knowledge, especially hidden knowledge (Holsapple (ed.), 2003; Cunningham et al., 2016). If the level of trust in an organization is high, its employees feel that sharing knowledge is safe and will not have negative consequences and that there is a mutual value that they will receive in the future or immediately in exchange for sharing their knowledge (Leistner, 2011, p. 65).

The third factor is management support. Management support signals to employees that sharing knowledge is important and encourages them to do so. Many studies have indicated that management support is related to the level of knowledge sharing among employees in an organization and without this support the degree of knowledge sharing is lower (Gupta, 2008; Tseng et al., 2012). Management board members, but also managers at all levels, must fully understand that the vast majority of knowledge is hidden in people's minds and that it cannot be "extracted" from the head and saved in some database. Being aware of this simple principle makes managers invest in people and the environment in which they operate. In an environment where people feel valued and have a high level of self-confidence, knowledge can flow more easily (Leistner, 2011, p. 68).

---

[5]These factors are related to the sterile organizational context indicated by Szulanski (1996) and to burdensome relations.

Competent leaders, supporting their teams and making them aware of the value of knowledge, is another important factor influencing the flow of knowledge. As leaders set an example to others working in the organization, they have a direct impact on the way the organization deals with and exchanges knowledge. Therefore, managerial leadership plays a key role in knowledge sharing and knowledge flows.

Knowledge flows in an organization will also be influenced by factors related to the structure and nature of the organization. The first of this group of factors is the size of the organization. It should be expected that in smaller organizations knowledge flows will be less formalized and more spontaneous than in larger entities. The type of knowledge that is transferred will also depend on the size of the organization—the larger the organization, the less tacit knowledge flows and more codified knowledge flows. This is related to the fact that in order to transfer tacit knowledge it is often necessary to have direct contact between people participating in the flow of knowledge, which is more difficult in a large organization. Moreover, large organizations focus more than smaller entities on the formal codification of employees' knowledge, acquiring their knowledge and making it available to other employees.

Another feature influencing the flow of knowledge is the hierarchical structure of an organization. A hierarchy is an effective mechanism for coordinating a complex system of many specialized units. Business enterprises are examples of hierarchies because they are composed of interconnected subsystems, each of which has a hierarchical structure down to the lowest level of the basic subsystem (Grant, 1996). With regard to knowledge flows, it can be expected that complex hierarchical structure of the organization will make them difficult. The hierarchical structure is related to a bureaucratic organization that relies heavily on rules and directives and the authority of the superiors. Today, most managers are aware of the disadvantages of bureaucratic structures that slow down processes and increase the constraints on knowledge flows. Moreover, such procedures are often time-consuming for knowledge to pass through every level of the organization (Al-Alawi et al., 2007). In a bureaucratic organization, tacit knowledge flows will be hampered, although tacit knowledge flows can be very complex and systematic. In an organization with a limited hierarchy and a flat structure, tacit knowledge flows will be freer because they will result from the expert approach of team members.

The type of activity of an organization that determines the approach to knowledge flows is also worth mentioning. Some organizations may not aim at blocking the flow of knowledge, but rather protecting them—for example, military organizations or organizations that carefully guard their organizational secrets.

Another aspect important for knowledge flows is the integration of knowledge flows with an organization's strategy. The integration of knowledge management activities with the strategy and goal of the organization is one of the factors influencing the success or failure of knowledge management (Wong, 2005). In the case of knowledge flows, it is also important to integrate them with the organization's strategy. Additionally, an important role is played by knowledge management strategy (if there is one in the organization), as well as functional strategies (marketing, personnel, etc.).

Knowledge flows are also determined by personal factors related to participants of knowledge flows. Such factors undoubtedly include the so-called background of people between whom knowledge flows, including their culture, religion, experience, values, personality type, etc. These factors will determine how willing these people will be to participate in the flows of knowledge as both sender and recipient. For example, it can be assumed that extroverted people will be more likely to participate in knowledge flows than introverts.

Knowledge flows will also depend on the presence of units to promote and support them. For example, in a small business where the owner frequently arranges meetings with employees and actively encourages them to exchange their knowledge and streamline their flows, knowledge will move more freely than in an organization where the owner will not pay any attention to it. Leaders can also act as facilitators of knowledge flows and intermediaries for knowledge flows. In the first case, leaders, aware that knowledge flows face barriers or are not efficient enough, can try to improve them, e.g. by encouraging employees to spend more time on knowledge flows or by solving technical problems related to the flows. In the second case, knowledge flow intermediaries become a link in the flow, connecting those who would normally not be able to participate in the flow of knowledge (e.g. due to the lack of awareness of the missing knowledge that someone in the organization has, or the inability to operate technical knowledge flow tools).

To sum up, a set of factors potentially influencing the flow of knowledge in organizations has been identified, which can be grouped into three categories: human factors, technical factors and organizational factors (Table 2.1).

2.8 Summary

In summary, in this chapter the concept of knowledge flows was introduced, together with the taxonomy of knowledge flows and a list of their potential determinants. There is a need for considering knowledge as a source of competitive advantage in the modern economy and organizations need to manage not only their own knowledge, but also the knowledge of customers. The concept of knowledge flows is a response to the current challenges, such as the need to obtain knowledge here and now (the so-called knowledge on request or even on-demand) or the increasing availability and universality of IT tools, with the simultaneous elimination of traditional forms of communication (e.g. individual conversations among employees). The description of knowledge-intensive business services and the companies offering them will be presented in the further part of the book. These companies will be the subject of research in the area of knowledge flows and knowledge management.

Table 2.1 Potential factors influencing the flow of knowledge

Type of factors	Description of the factor
Human factors	• function in the organization • sender and recipient motivation • recipient's absorption and retention capacity • trust • the so-called background of the sender and recipient (e.g. religion, work experience, personality type) • management support • leadership competencies • being seen as a credible source of knowledge • NIH syndrome • knowledge as a source of power and competence
Technical factors (including those related to the subject of knowledge flow)	• transmission channels (including IT infrastructure) • having time • having technical skills • ambiguity of knowledge • lack of knowledge verification • the value of knowledge
Organizational factors	• size of the organization • organizational context • organizational relations • organizational structure • organizational strategy and integration of knowledge flows into this strategy • type of activity of the organization • organizational culture • units promoting and supporting knowledge flows

References

Al-Alawi, A. I., Al-Marzooqi, N. Y., & Mohammed, Y. F. (2007). Organizational culture and knowledge sharing: Critical success factors. *Journal of Knowledge Management, 11*(2), 22–42. https://doi.org/10.1108/13673270710738898

Barney, J. (1991). Firm resources and sustained competitive advantage. *Journal of Management, 17* (1), 99–120.

Cohen, W. M., & Levinthal, D. A. (1990). Absorptive capacity: A new perspective on and innovation learning. *Administrative Science Quarterly, 35*(1), 128–152.

Cunningham, J., Seaman, C., & McGuire, D. (2016). Knowledge sharing in small family firms: A leadership perspective. *Journal of Family Business Strategy, 7*(1), 34–46. https://doi.org/10.1016/j.jfbs.2015.10.002

Currás, E. (2010). Ontologies, taxonomies and thesauri in systems science and systematics. In *Ontologies, taxonomies and thesauri in systems science and systematics*. Chandos Publishing. https://doi.org/10.1016/B978-1-84334-612-8.50001-6

Davenport, T. H., & Prusak, L. (1998). *Working knowledge: How organizations manage what they know*. Harvard Business Press.

DeCarolis, D. M., & Deeds, D. L. (1999). The impact of stocks and flows of organizational knowledge on firm performance: An empirical investigation of the biotechnology industry.

Strategic Management Journal, 20(10), 953–968. https://doi.org/10.1002/(SICI)1097-0266(199910)20:10<953::AID-SMJ59>3.0.CO;2-3

Dierickx, I., & Cool, K. (1989). Asset stock accumulation and the sustainability of competitive advantage: Reply. *Management Science, 35*(12), 1514–1514. https://doi.org/10.1287/mnsc.35.12.1514

Goh, S. C. (2002). Managing effective knowledge transfer: An integrative framework and some practice implications. *Journal of Knowledge Management, 6*(1), 23–30. https://doi.org/10.1108/13673270210417664

Grant, R. M. (1996). Toward a knowledge-based thoery of the firm. *Strategic Management Journal, 17*, 109–122.

Gupta, A. K., & Govindarajan, V. (2000). Knowledge flows within multinational corporations. *Strategic Management Journal, 496*(August 1999), 473–496. https://doi.org/10.1002/(sici)1097-0266(200004)21:4%3C473::aid-smj84%3E3.0.co;2-i

Gupta, S. K. (2008). A process-based classification of knowledge maps and application examples. *Knowledge and Process Management, 15*(1), 59–71. https://doi.org/10.1002/kpm

Holsapple, C. W. (Ed.). (2003). *Handbook on knowledge management 1. Knowledge matters*. Springer Verlag.

Hansen, M. T., Nohria, N., & Tierney, T. (1999). What's your strategy for managing knowledge? *Harvard Business Review, 72*(2), 106–116. https://doi.org/Article

Hsu, M.-H., & Chang, C.-M. (2014). Examining interpersonal trust as a facilitator and uncertainty as an inhibitor of intra-organisational knowledge sharing. *Information Systems Journal, 24*(2), 119–142. https://doi.org/10.1111/isj.12000

Inkpen, A. C., & Dinur, A. (1998). Knowledge management processes and international joint ventures. *Organization Science, 9*(4), 454–468. https://doi.org/10.1300/J042v05n01_10

Kowalczyk, A., & Nogalski, B. (2007). *Zarządzanie wiedzą. Koncepcja i narzędzia*. Difin.

Lambe, P. (2007). *Organising knowledge: Taxonomies, knowledge and organisational effectiveness*. Chandos Publishing.

Leistner, F. (2011). *Mastering organizational knowledge flow*. John Wiley & Sons.

Malamed, C. (2016). *Strategies for tacit knowledge transfer*. The E-Learning Coach. http://theelearningcoach.com/learning/tacit-knowledge-transfer/

McDermott, R., & O'Dell, C. (2001). Overcoming cultural barriers to sharing knowledge. *Journal of Knowledge Management, 5*, 76–85. https://doi.org/10.1108/13673270110384428

Nissen, M. E. (2005). Dynamic knowledge patterns to inform design: A field study of knowledge stocks and flows in an extreme organization. *Journal of Management Information Systems, 22*(3), 225–263. https://doi.org/10.2753/MIS0742-1222220308

Nissen, M. E. (2006). *Harnessing knowledge dynamics: Principled organizational knowing & learning*. IRM Press.

Nissen, M. E., & Bordetsky, A. (2011). Leveraging mobile network technologies to accelerate tacit knowledge flows across organisations and distances. In G. Trentin (Ed.), *Technology and knowledge flows: The power of networks* (pp. 1–25).

Pellini, A., & Jones, H. (2011). *Knowledge taxonomies - A literature review* (Issue May). http://www.odi.org.uk/resources/details.asp?id=5753&title=knowledge-taxonomies-literature-review&utm_source=ODI_Update&utm_medium=feed&utm_campaign=Google+Reader

Prusak, L. (2003). Accessed April 12, 2020, from http://www.gurteen.com/gurteen/gurteen.nsf/id/L001064/

Rogers, E. M. (1995). *Diffusion of innovations*. Free Press.

Sveiby, K. (2001). A knowledge-based theory of the firm to guide in strategy formulation. *Journal of Intellectual Capital, 2*(4), 344–358.

Szulanski, G. (1996). Impediments to the transfer of best practice within the firm. *Strategic Management Journal, 17*(1), 27–43. https://doi.org/10.1007/s13398-014-0173-7.2

Trentin, G. (Ed.). (2011). *Technology and knowledge flow. The power of networks*. Chandos Publishing. https://doi.org/10.1016/B978-1-84334-646-3.50001-0

Tseng, C. P., Chang, M. L., & Chen, C. W. (2012). Human factors of knowledge sharing intention among taiwanese enterprises: A preliminary study. *Human Factors and Ergonomics in Manufacturing, 22*(4), 328–339. https://doi.org/10.1002/hfm

von Hippel, E. (1999). "Sticky information" and the locus of problem solving: Implications for innovation. *Management Science, 40*(4).

Wong, K. Y. (2005). Critical success factors for implementing knowledge management in small and medium enterprises. *Industrial Management & Data Systems, 105*(3), 261–279. https://doi.org/10.1108/02635570510590101

Zhuge, H. (2002). Knowledge flow management for distributed team team software development. *Knowledge-Based Systems, 15*(8), 465–471. https://doi.org/10.1016/S0950-7051(02)00031-X

Zieba, M., Bolisani, E., & Scarso, E. (2016). Emergent approach to knowledge management by small companies: Multiple case-study research. *Journal of Knowledge Management, 20*(2), 292–307. https://doi.org/10.1108/JKM-07-2015-0271

Knowledge-Intensive Business Services 3

3.1 Knowledge-Intensive Business Services Sector: Features, Definitions, Classifications

The term "knowledge-intensive business services" (KIBS) was introduced in 1995 by the Miles team to describe services consisting in the creation, accumulation or dissemination of knowledge (Miles et al. 1995). According to researchers, services of this type are based on professional knowledge, and the companies that provide them are either a direct source of information/knowledge or use the knowledge to create indirect services for the production processes of their clients (Miles et al. 1995). Knowledge-intensive business services are typically targeted at other companies (both private and public) and can contribute to their competitive advantage by filling their knowledge gaps.

The interest in the sector of knowledge-intensive business services appeared in the 1990s, when for the first time the importance of this type of services in creating innovation and the development of economies was confirmed (European Commission, 2009; Tuominen & Toivonen, 2011). The development of the KIBS sector is seen as one of the characteristics of the knowledge-based economy (Aslesen & Isaksen, 2007a; Muller & Zenker, 2001b; Smedlund & Toivonen, 2007), and companies offering KIBS are referred to as organizations that are key to knowledge-based economies (Gadrey & Gallouj, 2002). KIBS companies perform a cross-domain role in the development of the knowledge economy by providing personalized, high-value business services to their clients (Wyrwich, 2013). The above rationale prompts a growing number of scientists and practitioners to study various aspects of the activities of companies offering knowledge-intensive business services, both at the macroeconomic and microeconomic level.[1] At the same time, in

[1]For example, the innovation of economies and regions (e.g. Aslesen & Isaksen, 2007a; Bettiol et al., 2012), innovation of companies offering KIBS (e.g. Amara et al., 2008; Toivonen & Tuominen, 2009), innovation of companies—clients and relations with them (e.g. Scarso &

M. Zieba, *Understanding Knowledge-Intensive Business Services*, Knowledge Management and Organizational Learning 10,
https://doi.org/10.1007/978-3-030-75618-5_3

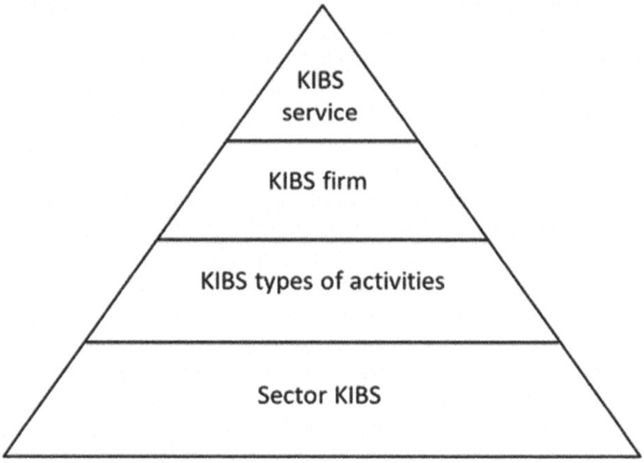

Fig. 3.1 Pyramid of definition issues related to knowledge-intensive business services (Source: Author's own illustration)

the literature, one can find many discrepancies regarding the definition of KIBS-type services, the companies offering them, as well as the sectors that can be included in this category of services. Definition issues can be presented in the form of a pyramid (Fig. 3.1). The narrowest category is the KIBS service (at the top of the pyramid), provided by KIBS companies. The next level of definition are types of activities classified as knowledge-intensive business services. The broadest category is the KIBS sector, which includes all companies offering knowledge-intensive business services. The following sections of this publication provide an overview of the definitions and classifications of the above terms based on the available literature.

The growing importance of knowledge-intensive business services in the economy reflects the need of various types of organizations for knowledge that helps them cope with changing technologies and social conditions. It is also the result of changes in organizational strategies and management methods, incl. The growing role of outsourcing, focusing on key competences and greater emphasis on intangible elements of the production of services and products (Miles, 2005, s. 39). The next two sections analyse two aspects of KIBS services—their knowledge intensity and business nature.

3.1.1 Features of Knowledge-Intensive Business Services

Knowledge-intensive business services have two distinguishing features—knowledge-intensive and focus on a business client, not an individual (Skórska, 2012, p. 97). *Knowledge intensity* is a term that is quite problematic both to define and to measure. This concept was created by analogy to other terms describing production factors, such as capital-intensive or labour-intensive. These terms indicate the

Bolisani, 2012; Shi et al., 2014), management in companies from the KIBS sector (e.g. Ritala et al., 2013; Sweeney et al., 2011; Zieba et al., 2016).

relative importance of labour, capital or knowledge in the production process (Skórska, 2012, p. 100). In a capital-intensive company, capital is more important than labour, and in a labour-intensive company, work is more important than capital. Similarly, in a knowledge-intensive company, this knowledge is more important than other production factors. However, unlike capital and labour intensity which are relatively easy to measure (Starbuck, 1992), knowledge intensity is a problematic issue, because the knowledge on which it is based is in itself a term difficult to define and measure, and many researchers define it in their own way. Despite these types of problems, attempts are made to define and measure the knowledge intensity that characterizes KIBS sector companies.

One of the indicators measuring the knowledge intensity of such companies is the structure of formal education of people working for such organizations (Koch & Stahlecker, 2006; Miles, 2005, 2011). However, although this indicator is relatively easy to measure, it has a serious imperfection: it does not take into account the education obtained outside formal training and the work experience of employees, which is often crucial for the activities of companies offering KIBS. Without specialized experience, often gained over the years, companies offering KIBS have fewer chances in competing against other similar entities. Moreover, the mentioned indicator does not take into account other forms of knowledge, such as, e.g., the knowledge possessed by the company in the form of databases and repositories, the organization's ability to learn or its ability to acquire knowledge from the broadly understood environment. Another drawback of this indicator is the fact that it measures the knowledge "at the entrance" in the company and does not take into account the knowledge "at the exit", e.g. in the form of product or service innovation (Koch & Stahlecker, 2006). The last of the mentioned imperfections of the indicator taking into account the structure of formal education of people working for KIBS companies does not occur in the case of measuring knowledge intensity by the amount of expenditure on research and development and the number of patents obtained by these companies (Koch & Stahlecker, 2006). However, information on patents may not be reliable as, due to the characteristics of KIBS companies, patent filing is relatively rare. A similar difficulty with credibility relates to R&D expenditure, which can be informal and non-measurable.

Knowledge intensity can also mean the non-routine nature of services (Muller & Doloreux, 2009). If this is the case, the meaning of "routine" should be defined. Here comes the question of how to define a routine. Is an action performed twice already routine or should it be done at least five times to deserve it? Perhaps the criterion of regular execution should be adopted, e.g. every day, at least once a week, etc. Another approach is to define knowledge intensity as the ability to integrate different sources of information and knowledge in innovative processes in the enterprise. According to this approach, companies offering KIBS are characterized by the ability to retrieve information from outside the company and transform this information, combined with knowledge about the company—the client, into services useful for their clients (Hipp, 1999). Another approach was proposed by Starbuck, according to whom the knowledge-consuming nature of a company is demonstrated by having unique and valuable experience, and not popular and commonly available knowledge. It should be remembered that having expert knowledge should be considered in the context of the company's effects—they determine whether the company is knowledge-intensive and whether it is able to make appropriate use of its expertise (Starbuck, 1992).

In conclusion, there are several approaches to the issue of what the knowledge intensity of KIBS companies is. Some of these approaches use indicators that are easier to measure (e.g. the formal employment structure of people working in these companies), others—more difficult (e.g. determining the degree of routine service offered). According to the author of this work *it is reasonable to treat the knowledge usage of KIBS companies as having, using and improving the skills of acquiring and processing specialist knowledge necessary for the functioning of other organizations, providing products/services to their clients and achieving the intended goals.* While measuring the knowledge intensity determined in this way may be problematic, it can be assumed that in qualitative research this approach is applicable, and the KIBS sector research at the quantitative level is always based on a predefined population of selected sectors. Therefore, this book uses the proposed approach developed for the needs of qualitative research.

Another important feature of the KIBS sector services, indicated by many researchers, is that they are addressed mainly to companies or organizations, and not to individuals or households. (Toivonen, 2004). Knowledge-intensive business services are largely based on professional (expert) knowledge related to a given field (den Hertog, 2000).

Due to the fact that KIBS companies offer services with a high degree of tailoring to the needs of their clients, the creation of such services requires close and intensive cooperation between the KIBS sector company and its clients. Without customer cooperation with a company that offers KIBS, it is impossible to obtain tacit knowledge residing in the customer's organization, which is an important element of knowledge-based services. Some authors even refer to the term co-production, emphasizing the role of the client in creating any service (e.g. Büttgen & Ates, 2009; Spohrer & Maglio, 2008), and in the case of knowledge-intensive services, it can be assumed that due to its specialized nature, this co-production is indispensable. This is confirmed by the research cited by Miles (2008), stating that KIBS service providers rely on knowledge gained from companies—customers to a greater extent than manufacturing companies and most other service companies. In other words, the KIBS service cannot be properly created and delivered without detailed information on the client's needs; knowing about a similar customer is usually not enough (Doroshenko et al., 2013). Customers and users of services often have to be somehow involved in the production process of the service. Sometimes physical presence is enough, but often not only the client's input is required, but also intense interaction with the service provider and sometimes with other clients as well (Doroshenko, 2012; Miles, 2012). Summarizing, it can be stated that *the business nature of KIBS services is not only that they are addressed to business entities, not physical entities, but also that they are created in close cooperation with the client.*

3.1.2 Definitions of Knowledge-Intensive Business Services and the Companies Offering Them

The two main features of KIBS services presented above allow us to move to the definition of knowledge-intensive business services. Since I. Miles and his colleagues introduced this term into the literature in 1995, various researchers have been trying to define what these services really are. Not only are their definitions created, but also attempts are made to define their characteristic features or assign particular types of activity to the KIBS sector. Table 3.1 presents the most important definitions and features of business knowledge-intensive services presented in the literature.

As can be seen (Table 3.1), individual researchers pay attention to various aspects of business knowledge-intensive services. Miles et al., who were the first to define this concept, drew attention to knowledge-related processes such as creation and dissemination. In the same year, O'Farrell and Moffat defined business services in general (including the category of knowledge-intensive companies) as "services that provide clients with strategic information and knowledge that is relatively intangible, potentially sustainable in its effects, and related to problem solving and strategy development, not routine problems" (O'Farrell & Moffat, 1995). This definition also applies to knowledge-intensive business services as it indicates strategic information and knowledge provided by service companies, and knowledge-intensive business services are considered in it as a subset of business services. Among other important aspects of KIBS services, resulting from the available descriptions, it can be noted that they are provided not only to other companies, but also to public organizations

Table 3.1 Definitions of knowledge-intensive business services according to various authors

Definition of knowledge-intensive business services	Author
Services that lead to the creation, accumulation and dissemination of knowledge	Miles et al. (1995)
Services provided by companies to other companies or public sector entities where experience and skills play a particularly important role	Toivonen (2004)
Highly application services where tacit knowledge plays an important role	Koch and Strotmann (2005a)
Personalized services with a wide scope and high level of security	Pardos et al. (2007)
Services that are non-routine	Muller and Doloreux (2009)
Business-to-business (B2B) services based on expert knowledge, where knowledge plays an important role both in the process of providing the service and as a result of the service. Customer problems often have some unique features that prevent KIBS companies from unifying their knowledge base: Learning and searching for new knowledge is in many cases part of the KIBS service delivery process (p. 394)	Tuominen and Toivonen (2011)
Services where knowledge is the main element in both entry and exit	Desmarchelier et al. (2013)

Source: Own elaboration, based on the mentioned resources

Table 3.2 Characteristics of knowledge-intensive business services according to various authors

Characteristics of knowledge-intensive business services	Author
They rely heavily on professional/specialist knowledge; they are the primary source of knowledge/information or use knowledge to create indirect services in the production processes of their clients; they are important for the competitiveness of companies and are addressed mainly to them	Miles et al. (1995)
They are based on knowledge or experience in a specific technical or operational field; are intermediate products	den Hertog (2000)
They require specialized knowledge and learning processes	Koch and Strotmann (2008)
Typically, the service provider and the customer engage in intense interaction that creates and shares new knowledge	Tuominen and Toivonen (2011)

Source: Own elaboration, on the basis of provided resources

(Toivonen, 2004). Usually, the KIBS sector does not include services provided to households or natural persons. KIBS services are also known as highly personalized (Tuominen & Toivonen, 2011) and non-routine (Muller & Doloreux, 2009). Researchers who define KIBS services also point to the great importance of knowledge at every stage of preparing, creating and delivering the KIBS service (Desmarchelier et al., 2013).

In addition to defining the KIBS service, some researchers present the characteristics of the KIBS sector service (Table 3.2), emphasizing the high degree of dependence on specialist, technical and operational knowledge as well as experience gained (den Hertog, 2000; Koch & Strotmann, 2008; Miles et al. 1995). Attention is also drawn to the fact that these services are often intermediate products in the sense that companies—clients use them in their production processes (den Hertog, 2000; Miles et al. 1995). In addition, as mentioned in the previous section, customers often need to work closely together to develop these services by sharing knowledge and creating new (Tuominen & Toivonen, 2011).

Summarizing both the review of the definitions and the characteristics of KIBS-type services, the following definition of this type of service is proposed:

- *Knowledge-intensive business services are services addressed to business entities (enterprises and various types of organizations), which are largely based on specialist knowledge and experience, are often non-routine, tailored to the client's needs, and sometimes even created in close cooperation with the client.*

These two main features of knowledge-intensive business services—knowledge-intensive and business-related—were used to define them (Fig. 3.2).

Companies offering knowledge-intensive business services differ from production companies not only in terms of the subject of production, but also in its nature. The products of production companies contain a high level of codified knowledge (they are perceived as "commodification of knowledge"), and the services of KIBS sector companies contain a lot of intangible assets and tacit knowledge (Muller &

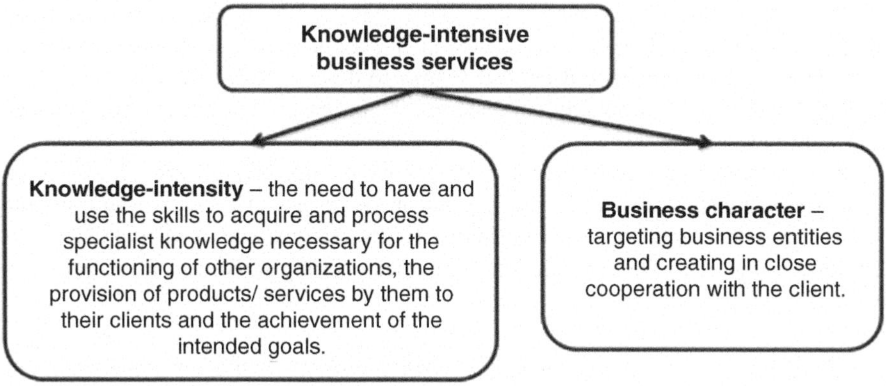

Fig. 3.2 Definition of knowledge-intensive business services through the prism of their knowledge intensity and business nature (Source: Author's own illustration)

Doloreux, 2009, p. 67). KIBS sector companies have common features, and at the same time are heterogeneous. Strambach distinguishes three main features that combine heterogeneous KIBS subsectors into one sector:

- Knowledge is not only a key factor in the production of these companies, but it is also the good they sell. In most cases, these companies provide intangible services that are largely knowledge-based.
- The delivery of knowledge-intensive services requires in-depth interaction between the provider and the user, and both parties are involved in cumulative learning processes. Using knowledge-intensive services cannot simply be equated with the purchase of standard external services (outsourcing).
- Advisory activity, understood as a problem-solving or concept development process, in which KIBS companies match their experience and expertise to customer needs, is based on interactions between KIBS companies and their clients (Strambach, 2010, p. 173).

In turn, den Hertog (2000) and Miles et al. (1995) pay attention to the important aspect of innovation of the KIBS sector companies, which are not only innovative in themselves, but also help other companies in their innovation processes. The characteristics of KIBS companies also include the aspect of matching the service to the customer's needs or solving a specific problem faced by the customer (Bettencourt et al., 2002; Herstad & Ebersberger, 2014; Koschatzky & Stahlecker, 2006; Miles, 2005). Interestingly, companies from the KIBS sector are identified by some researchers with consulting companies in the broad sense of the word (Consoli & Elche-Hortelano, 2010; Muller, 2001; Muller & Zenker, 2001a). This is a simplification: although every consulting company is a company from the KIBS sector, because in cooperation with the client, it provides the missing knowledge, there are also companies that are not consulting companies, but belong to the KIBS sector (e.g. design offices, law firms or advertising agencies). Another feature of the

KIBS sector companies, pointed out by Toivonen (2004), is the creation by these companies of a network that connects customers, cooperators, public institutions and research and development centres. It is thanks to this network that companies from the KIBS sector can more effectively acquire the necessary knowledge and generate new solutions.

On the basis of the analysis of the descriptions of companies offering knowledge-intensive business services (Table 3.3), it can be seen that these descriptions are clearly consistent with the descriptions of KIBS services and their characteristics (Tables 3.1 and 3.2). Descriptions of companies offering KIBS services very often include:

- knowledge as the basic source of creating and delivering services by these companies;
- the ability to acquire and integrate various types of information and knowledge potentially useful for the client and helping to solve his problems;
- element of non-routine nature;
- ability to create networks and co-create innovations in companies—clients.

Based on the descriptions presented in Table 3.3 it can also be inferred that KIBS companies were initially seen only as providers of specialized knowledge and information to their customers. With time, it was noticed that the information and knowledge provided by these companies are often the result of mutual interactions or even co-productions with companies—clients. In conclusion, it can be stated that there is no single, commonly used definition of KIBS services and companies offering these services. Wood came to similar conclusions (Wood, 2002), the same as Tether and Tajar (2008). However, you can indicate the following aspects appearing in individual descriptions and on their basis characterize this type of services and the companies offering them:

- Knowledge is a key element in creating a business knowledge-intensive service and is a key resource in a company offering such services.
- The KIBS service consists mainly in the transfer of knowledge and skills possessed by KIBS to the client's organization.
- The KIBS service is produced in close cooperation with the company—the client, and often also with other entities, e.g. research and development units.
- Knowledge flow in the process of creating a KIBS service is a very important element and a success factor.
- Companies providing KIBS services combine a variety of knowledge and information to solve their client's specific problem.

KIBS companies interact with their broadly understood environment (partners, competitors, research and development institutions, etc.) and as a result of this interaction they create typical knowledge processes, such as: absorption, analysis, codification and integration (Fig. 3.3). The knowledge created in these processes is

Table 3.3 Companies offering KIBS by various authors

Characteristics of companies offering KIBS	Author
Companies that are users, carriers and sources of innovation	Miles et al. (1995)
Companies characterized by the ability to receive information from outside the company and transform this information together with knowledge in the company into specialized services for their clients	Hipp (1999)
Companies that are co-creators of innovations in their clients' companies	den Hertog (2000)
Consulting companies in a broader sense Companies that provide services with high intellectual added value, mainly for other companies	Muller and Zenker (2001a) Muller (2001), s. 2
Enterprises whose core value-added activity is the accumulation, creation and dissemination of knowledge in order to develop a solution (tailored service or product) that meets the customer's needs	Bettencourt et al. (2002), s. 100–101
A company that is a knowledge accumulation system that collects, benefits and protects knowledge from various trading services	Gadrey and Gallouj (2002)
Companies that have numerous and comprehensive contacts with various entities and create a network connecting clients, cooperators, public institutions and research and development centres	Toivonen (2004)
Companies locating, developing, combining and implementing various types of technological knowledge and applying it to solve specific problems of their clients. These companies are involved in the fusion of local knowledge	Miles (2005)
For these companies, knowledge is not only the main means of production, but also the product they sell	Simmie and Strambach (2006)
Companies combining different types of highly specialized knowledge to create (both innovative and non-innovative) solutions to specific problems	Koschatzky and Stahlecker (2006)
Companies that provide information, new technologies, new production/sales techniques, new ways of managing to new markets	Pardos et al. (2007)
Service companies whose services, aimed at other companies and organizations, are characterized by a high intensity of knowledge	Muller and Doloreux (2009)
Intermediary companies that specialize in knowledge screening, assessment and evaluation and provide professional consulting services	Consoli and Elche-Hortelano (2010)
Companies that are characterized by a high intensity of knowledge and provide their clients mainly with non-routine services	Doloreux and Shearmur (2010)
Companies "producing" knowledge that cannot be "produced" in the company—The client (e.g. due to cost issues)	Doroshenko (2012)
Firms that specialize in creating, verifying and applying knowledge to solve their clients' problems	Herstad and Ebersberger (2014)

Source: Own elaboration

offered to companies—clients and further developed as a result of close cooperation between KIBS and the company—client.

It can be said that this approach is related to the approach represented by Hipp (1999), which named KIBS companies as intermediaries between knowledge producers and its users. Then, as a result of the processes and interactions described,

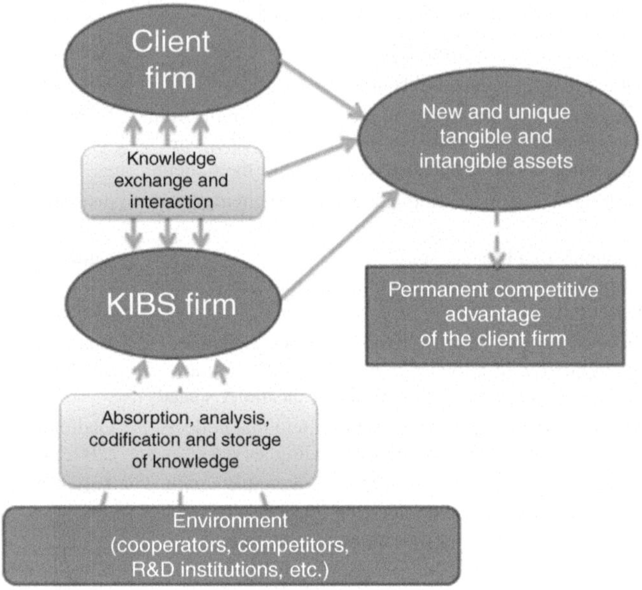

Fig. 3.3 The role and activity of KIBS companies in the environment (Source: Author's own illustration)

new and unique tangible and intangible assets are created. These assets may take the form of innovations or modified solutions in companies—clients. They are also often a response to a specific problem faced by a company—a customer. Regardless of their form, these tangible/intangible assets are a potential source of sustainable competitive advantage in companies—clients. *The desired effect of the provision of services by KIBS companies is a permanent competitive advantage of companies—customers.*

3.1.3 Services Included in the Sector of Knowledge-Intensive Business Services

On the basis of the previously presented definitions and descriptions of various elements of knowledge-intensive business services, it can be determined what kind of services they are and what their place in the service sector. Due to the fact that using the presented definitions of knowledge-intensive companies at the macroeconomic level is not an easy task, it is necessary to define the types of services that companies from the KIBS sector deal with. As in the case of various definitions, also in relation to the services included in KIBS, researchers are not unanimous about which types of economic activity should be classified as knowledge-intensive business services (Table 3.4).

Table 3.4 Services included in the KIBS sector according to selected authors

KIBS services	Author
Data processing, consulting in the field of computer hardware and software, management consulting, tax services, audit services, technical and engineering, advertising, marketing and communication services and consulting	Strambach (2001)
Computer and related, research and development, legal, accounting, auditing, market and public opinion research, business and management consultancy, architectural and engineering activities, technical research and analysis, advertising, recruitment and provision of employees, photography activities, education and training services	Toivonen (2004)
Computer and related activities, research and development, accounting and market research, market and public opinion research, business and management consulting, architectural and engineering activities, technical research and analysis, advertising, recruitment and provision of employees, photography activities, commercial activity remained	Muller and Doloreux (2009)

A large number of researchers, as well as most statistical offices of OECD countries, use official industry classifications to define the list of subsectors belonging to the KIBS sector. According to Freel (2006), although these lists are still characterized by a certain ambiguity and interpretive subjectivity, they provide a good basis for research into the KIBS sector. The most popular classification used in research on the KIBS sector in Europe is the NACE classification (NACE Rev. 1.1., NACE 2), while the vast majority of publications on the KIBS sector are based on the NACE Rev.1 classification (Miles, 2011). Individual researchers dealing with the subject of the KIBS sector present partially divergent opinions regarding the departments belonging to the KIBS sector, separated on the basis of the NACE classification. According to Baláž (2004a) the KIBS sector includes a wide range of services, namely NACE Rev.1 divisions 64–67 and 70–74. This author emphasizes the special importance of knowledge-intensive information and communication business services that enhance the transfer of KIBS to all users in the national economy, thus enabling the spread of KIBS in other sectors of the economy. Thanks to this, technological and organizational innovations arise and are applied. A much narrower approach to the KIBS sector is presented, for example, by Miles (2005), which includes in the KIBS sector only divisions 72–74 of NACE Rev.1, i.e. computer services (division 72), research and development services (division 73) and other business services, such as accounting, consulting, legal, engineering and advertising services (division 74), excluding some groups.[2] In his later publication, Miles (2011) excluded from the KIBS sector groups: 74.5, 74.81 and 74.82. Other researchers have also changed their minds over the years, including other divisions and groups of the NACE classification in the KIBS sector. For example, Muller and Zenker (2001a) in 2001, they included only two subsectors in the KIBS

[2]Miles excluded the following groups from the KIBS sector: 74.6 (investigation and security activities); 74.7 (cleaning and cleaning of facilities); 74.82 (packaging activities); 74.83 (translation and secretarial services) (Miles, 2005).

sector: 72 and 74, and in 2009 they added the 73 subsector. An overview of the categories included in the KIBS sector is presented in Table 3.5.

As it can be seen (Table 3.5), the researchers mostly agree to include in the KIBS sector division 72 (computer and related services), division 73[3] (research and development activity) and selected groups of division 74 (other economic activity), namely groups: 74.1, 74.2, 74.3, 74.4. As for the remaining departments and groups, the researchers disagree and some of them present the KIBS sector in a broader sense (e.g. Baláž, 2004a, 2004b), and others include or exclude from it, for example, other economic activities or photographic activities (e.g. Huggins, 2011; Muller & Zenker, 2001a).

In connection with the introduction in 2008 of the modified NACE Rev.2 classification, the researchers attempted to verify the subsectors belonging to the KIBS sector. Schnabl and Zenker (2013) proposed to include in the KIBS sector the divisions 62, 63, 69, 70, 71, 72 and 73 of the NACE Rev.2.0 classification (Table 3.6).

Despite the introduction of the NACE Rev.2.0 classification already in 2008, still few publications are based on the separation of the KIBS sector based on this classification (e.g. Ferreira et al., 2013). In one of the publications of this type, researchers classified the KIBS sector in the following subclasses: 62.01, 62.02, 62.03, 62.09, 63.11, 63.91, 63.99, 69.10, 69.20, 71.11, 70.22, 71.12, 71.20, 72.1, 72.2, 73.11, 73.12, 73.20, 74.20, 74.90, 78.10, 78.30[4] (Fernandes et al., 2013). These authors included in the KIBS sector, inter alia, photographic activities and activities related to searching for jobs and acquiring employees about which researchers do not agree whether they belong to the KIBS sector or not. Although publications based on the NACE Rev.2.0 classification appear, many works from recent years are still based on the previous versions of the NACE classification (Rev. 1.0 and Rev. 1.1.) (Bolisani et al., 2014; Ciriaci et al., 2015; Ciriaci & Palma, 2016; Deza & Lopez, 2014). This is due to from the availability of macroeconomic data collected in accordance with the previously applicable NACE Rev.1.1 or Rev.1.0 classification and very deeply rooted in the literature, the KIBS sector separation

[3]Only two articles are an exception here: Muller (2001) and Muller and Zenker (2001a).

[4]62.01—Activities related to software; 62.02—Activities related to IT consultancy; 62.03—Activities related to the management of IT devices; 62.09—Other service activities in the field of information technology; 63.11—Data processing; website management (hosting) and similar activities; 63.91—Activities of news agencies; 63.99—Other information service activities, not elsewhere classified; 69.10—Legal activities; 69.20—Accounting and bookkeeping activities; tax consultancy; 71.11—Activities in the field of architecture; 70.22—Other business and management consultancy; 71.12—Engineering activities and related technical consultancy; 71.20—Research and analysis relating to food quality; 72.1—Research and development in the field of biotechnology, Scientific research and development in the field of other natural and technical sciences; 72.2—Research and development in the field of social sciences and humanities; 73.11—Advertising agencies; 73.12—Brokerage in the sale of time and space for advertising purposes; 73.20—Market research and public opinion research; 74.20—Photography activities; 74.90—Other professional, scientific and technical activities, not elsewhere classified; 78.10—Activities relating to job search and recruitment; 78.30—Other activities related to the provision of employees.

Table 3.5 KIBS sector—overview of subsectors included

NACE Rev.1.1	Description	Muller (2001) and Muller and Zenker (2001a)	Baláž (2004a)	Miles (2005)	Miozzo and Grimshaw (2006)	Koch and Strotmann (2008)	Muller and Doloreux (2009)	Huggins (2011)	Miles (2011)	Skórska (2012)
64	Postal and telecommunications services		X							
65	Financial intermediation, except insurance and pension funding		X							
66	Insurance and pension funds, without legally guaranteed social welfare		X							
67	Activities auxiliary to financial intermediation and insurance		X							
70	Real estate services		X							
71	Renting of machines and devices without operator and of personal and household goods		X							
72	Computer and related services									
72.1	Consultancy in the field of computer hardware	X	X	X	X	X	X	X	X	X
72.2	Software activities	X	X	X	X	X	X	X	X	X
72.3	Data processing	X	X	X	X	X	X	X	X	X
72.4	Database activities	X	X	X	X	X	X	X	X	X

(continued)

Table 3.5 (continued)

NACE Rev.1.1	Description	Muller (2001) and Muller and Zenker (2001a)	Baláž (2004a)	Miles (2005)	Miozzo and Grimshaw (2006)	Koch and Strotmann (2008)	Muller and Doloreux (2009)	Huggins (2011)	Miles (2011)	Skórska (2012)
72.5	Maintenance and repair of office machinery, accounting and computer hardware	X	X	X	X	X	X	X	X	
72.6	Activities related to IT, other	X	X	X	X	X	X	X	X	X
73	Research and development activities									
73.1	Research and development in the field of natural and technical sciences		X	X	X	X	X	X	X	X
73.2	Research and development in the field of social sciences and humanities		X	X	X	X	X	X	X	X
74	The remaining economic activity									
74.1	Legal, accounting and bookkeeping activities; counselling; management of holdings	X	X	X	X	X	X	X	X	X
74.11	Legal activities	X	X	X	X	X	X	X	X	X
74.12	Accounting and bookkeeping activities	X	X	X	X	X	X	X	X	X

74.13	Market research and public opinion research	X	X	X	X	X	X	X	X	X
74.14	Business and management consultancy	X	X	X	X	X	X	X	X	X
74.2	Activities in the field of architecture and engineering	X	X	X	X	X	X	X	X	X
74.3	Technical research and analysis	X	X	X	X	X	X	X	X	X
74.4	Advertising	X	X	X	X	X	X	X	X	X
74.5	Recruiting and sharing employees		X	X	X			X		X
74.6	Detective and security activities		X		X					
74.7	Cleaning of facilities		X		X					
74.8	Economic activity not elsewhere classified		X	X	X					
74.81	Photography activities		X	X	X					
74.82	Packaging activities		X		X					
74.83	Translation and secretarial services		X		X					
74.87/ 74.84	The remaining economic activity	X	X	X	X		X			

Table 3.6 KIBS sector in the NACE Rev.2.0 classification

Sector KIBS NACE Rev.2	Group name	Class description
Section J, division 62	Information and communication	Activities related to computer programming, consultancy and related activities
Section J, division 63	Information and communication	Information service activities
Section M, division 69	Professional, scientific and technical activity	Legal, accounting and tax consulting activities
Section M, division 70	Professional, scientific and technical activity	Activities of central companies, management consultancy
Section M, division 71	Professional, scientific and technical activity	Activities in the field of architecture and engineering Technical research and analysis
Section M, division 72	Professional, scientific and technical activity	Research and development work
Section M, division 73	Professional, scientific and technical activity	Advertisement Market research and public opinion research

Source: Schnabl and Zenker (2013), p. 5

based on previous versions of the NACE classification. An increasing number of publications based on NACE Rev.2.0 or its equivalents can be expected, but it will most likely be a gradual and slow process.

As it can be seen, separating KIBS companies based on official industrial classifications is a popular practice among researchers in this sector. Meanwhile, some researchers try to define the KIBS sector in a different way, justifying it with the fact that official industrial classifications are not perfect, and in more than 50% of cases, the selected official area of economic activity does not correspond to the actual activities of companies (Hipp, 1999). That is why Hipp (1999) proposed the conceptualization of KIBS companies in relation to the creation of knowledge by these companies (Fig. 3.4).

When considering the use of this type of approach, however, it should be borne in mind that using official industrial classifications greatly facilitates the identification and reaching of KIBS companies. Checking to what extent a given company meets the criteria proposed by Hipp (1999) would be a very laborious activity and practically impossible to implement in the case of conducting research on a large sample. Therefore, despite the obvious imperfections of the official industrial classifications, using them seems to be a necessity, especially when conducting quantitative research.

In this study, the following divisions of the NACE Rev.1.1 classification were included in the KIBS sector:

- 72.1 Consultancy in the field of computer hardware;
- 72.2 Software activities;
- 72.3 data processing;
- 72.4 Database activities;

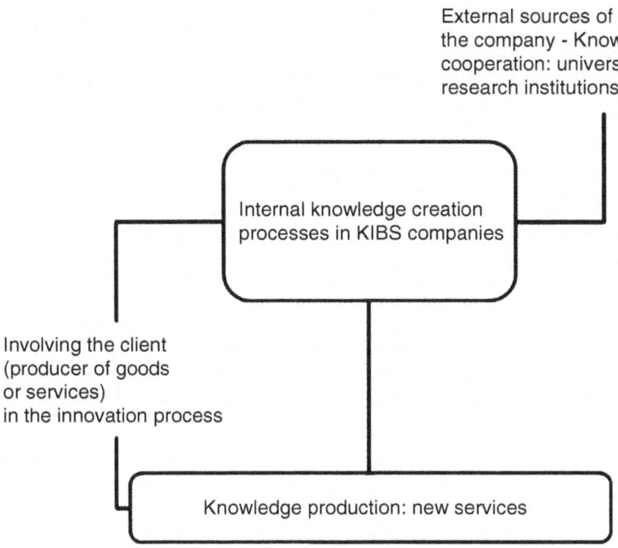

Fig. 3.4 Conceptualization of knowledge-intensive business services according to Hipp (1999) (Source: Author's own illustration, based on: Hipp, 1999)

- 72.5 Maintenance and repair of office machinery, accounting machines and computer hardware;
- 72.6 Other activities related to computer science;
- 73.1 Research and development in the field of natural and technical sciences;
- 73.2 Research and development in the field of social sciences and humanities;
- 74.1 Legal, accounting and bookkeeping activities; market research and public opinion polling, business and management consultancy; activities of holding companies;
- 74.2 Architectural and engineering activities;
- 74.3 Technical research and analysis;
- 74.4 Advertising.

Based on the study (Schnabl & Zenker, 2013) the KIBS sector includes the following divisions of the NACE Rev.2.0 classification: 62, 63, 69, 70, 71, 72 and 73.

According to the author, in the case of companies conducting the above-mentioned types of activity, there is no doubt about both the knowledge-intensive nature and the business nature of the services they offer. Thus, using the criterion of meeting the definition of a knowledge-intensive business service proposed in the Sect. 3.1.2, these types of activity were selected as belonging to the KIBS sector. It is worth noting that these activities are included in the KIBS sector by some researchers (e.g. Koch & Strotmann, 2008; Miles, 2011; Muller & Doloreux, 2009). After defining the KIBS sector, the next section will present the different divisions of this sector used in the literature.

3.1.4 Classifications of Knowledge-Intensive Business Services

Due to the heterogeneity of the KIBS sector, researchers are trying to divide it into smaller subgroups, guided by the criteria of their choice. The first typology of the KIBS sector was proposed by Miles et al. and divided the KIBS sector into two subgroups: the so-called traditional consulting services (KIBS I), using new technologies, and new knowledge-intensive services (KIBS II), creating these technologies (Miles et al. 1995). The first group included, among others marketing and accounting services, while the latter—e.g. software design or technical engineering services (Table 3.7).

Over time, the division into KIBS I and KIBS II was replaced by a different division: into P-KIBS (i.e. professional KIBS services) and T-KIBS (i.e. KIBS technical services) (den Hertog, 2000; Miles, 2008; Nählinder, 2005; Nählinder & Hommen, 2002). KIBS professional services are based on specialist knowledge of administrative systems and social relations (e.g. management services, accounting and legal services, market research services, etc.). In turn, the technical services of KIBS are based mainly on information and communication technologies, as well as on the production and transfer of knowledge related to technology (e.g. computer services, research and development services, engineering services, etc.) (Muller & Doloreux, 2009). Although all companies from the KIBS sector very often employ university graduates, T-KIBS companies more often hire graduates of science and engineering (S&E), and P-KIBS companies—graduates of other fields of study

Table 3.7 The typology of KIBS services according to Miles et al. (1995)

KIBS I—traditional consultancy services that make intensive use of new technologies	KIBS II—new knowledge-intensive business services based on technology
Marketing/advertising Training (apart from training in the field of new technologies) Designing (apart from design using new technologies) Certain financial services (e.g. securities trading and other activities related to the stock and securities market) Other services (except services requiring new office equipment and "physical" services such as cleaning) Construction services (e.g. architecture, expertise, structural engineering, but excluding services using new IT equipment) Management consulting (apart from consulting based on new technologies) Bookkeeping and accounting Legal services Services related to environmental protection (not requiring new technologies, e.g. environmental law)	Computer networks/telematics (e.g. online databases) Certain telecommunications services (especially new business services) Software Other computer services (e.g. hardware management) Training in the field of new technologies Design requiring new technologies Office services requiring new office equipment Construction services (mainly requiring new IT equipment) Management consulting that requires new technologies Technical engineering Environmental services requiring new technologies, e.g. monitoring, laboratory/ science services R&D consulting and "high-tech agencies"

Source: Miles et al. (1995), pp. 29–30

(Miles, 2011). The division into T-KIBS and P-KIBS is partially divergent in the case of individual authors. For example, Nählinder and Hommen (2002) they did not include subsector 72.5 (maintenance and repair of office machinery, accounting machines and computer equipment) in the T-KIBS category, which was included in this category by other researchers (Koch & Strotmann, 2004; Koch & Stahlecker, 2006). In addition, some subsectors were placed in opposite categories, e.g. 73.2 (R&D in social sciences and humanities). The division into T-KIBS and P-KIBS according to selected authors is presented in Table 3.8.

Another subcategory of the KIBS sector was distinguished by Skogli (1998), who proposed the technology-related KIBS subsector rather than technology-based, as proposed by Miles et al. (1995). Skogli included the following groups of the NACE Rev.1 classification in this subsector: computer and IT services (division 72); research and development works (division 73)—only the private part of this sector; architectural and engineering activities (group 74.2); technical testing and analysis (group 74.3) (Skogli, 1998).

Another subcategory of KIBS services was proposed in 2011 by Miles—these are services referred to as C-KIBS (creative KIBS, i.e. creative services of KIBS) (Miles, 2011). These services are related to cultural and similar knowledge, and require aesthetic skills and creativity. These include services such as advertising, graphic and other design and media services. Some C-KIBS services have been included in the T-KIBS or P-KIBS categories (e.g. marketing research or architectural services), but others have been included in the category "services not elsewhere classified" (Miles, 2012, p. 22).

On the other hand, in another study you can find a division into three subcategories of KIBS companies:

- *technological KIBS*: research and development services in the field of life sciences, medicine, engineering and technology, activities in the field of civil and construction engineering, electrical engineering, designing machine building and technology, other activities in the field of architecture, engineering, technical consultancy and technical research and analysis;
- *IT KIBS*: consultancy in the field of computer hardware and software, data processing and databases;
- *other non-technological KIBS*: services in the field of legal advisory, legal representation, consulting related to patents, other legal activities, audit services, market and public opinion research, consulting in the field of business management and business, urban planning, advertising agencies, translation services and industrial design (Werner, 2001).

An even larger number of subcategories of the KIBS sector, based on the NACE and ISIC Rev.3 classification groups, were distinguished by Toivonen (2004). The author proposed nine subgroups of KIBS companies (Table 3.9).

Interestingly, Toivonen (2004) it was the only one to include selected groups from division 80 (education) in the KIBS sector. However, one can doubt whether the broadly understood education actually belongs to the sector of

Table 3.8 Division into T-KIBS and P-KIBS according to selected authors

KIBS category	T-KIBS		P-KIBS	
Study	Nählinder and Hommen (2002)	Koch and Strotmann (2004) and, Koch and Stahlecker (2006)	Nählinder and Hommen (2002)	Koch and Strotmann (2004) and Koch and Stahlecker (2006)
72.1 consultancy in the field of computer hardware	X	X		
72.2 software activities	X	X		
72.3 data processing	X	X		
72.4 database activities	X	X		
72.5 maintenance and repair of office machinery, accounting and computer equipment		X		
72.6 other activities related to computer science	X	X		
73.1 research and development in the field of natural and technical sciences	X	X		
73.2 research and development in the field of social sciences and humanities	X			X
74.1 legal, accounting and bookkeeping activities; counselling; management of holdings			X	X
74.2 architectural and engineering activities	X	X		
74.3 technical research and analysis	X	X		
74.4 advertising			X	X
74.5 recruitment and provision of employees			X	
74.90 other professional, scientific and technical activities not elsewhere classified			X	

Table 3.9 KIBS subcategories according to Toivonen (2004)

Subcategory	NACE	Group NACE	Description of economic activity
1	72		*Computer and related services*
		72.1 72.2 72.3 72.4 72.6	Consultancy in the field of computer hardware Software consultancy Data processing Database activities Other activities related to IT
2	73		*Research and development activities*
		73.101 73.102 73.103 73.200	Research and development work in the field of medical sciences and pharmacy Research and development works in the area of other environmental sciences Research and development work in the field of technical sciences Research and development in the field of social sciences and humanities
3	74		*The remaining economic activity*
		74.111– 74.119 74.111 74.112 74.113 74.119	Legal activities Legal advice Activities in the field of legal representation Patent consultancy Other legal action
4	74		*The remaining economic activity*
		74.121– 74.129 74.121 74.122 74.129 74.842	Accounting and bookkeeping activities Accounting services Audit services Other accounting services Debt collection services
5	74		*The remaining economic activity*
		74.401– 74.409 74.130 74.401 74.402 74.409 74.843	Advertising services Market research and opinion polls Activities of advertising agencies Direct and external advertising activities Other advertising activity Activities related to the organization of fairs and exhibitions
6	74		*The remaining economic activity*
		74.201– 74.209 74.300 74.841	Activities in the field of architecture and engineering
		74.201 74.202 74.203 74.204 74.205	Urban spatial planning Construction design activities Activities in the field of architecture Engineering activities Design of heating, pipelines and air conditioning

(continued)

Table 3.9 (continued)

Subcategory	NACE	Group NACE	Description of economic activity
		74.206 74.207 74.208 74.209 74.300 74.841	Design in the field of energy Other design activity Mechanical and process design Other architectural and engineering activities and technical consultations Technical research and analysis Industrial design
7	74		*The remaining economic activity*
		74.140	Business and management consultancy
8	75		*The remaining economic activity*
		74.501– 74.509 74.501 74.502 74.509	Recruiting and sharing employees Recruitment activities Activities related to the provision of employees Other activities related to work and employees
9	80		*Education*
		80.220 80.300 80.421– 80.429	Education in the private sector
		80.220 80.300 80.421 80.422 80.423 80.424 80.429	Technical and vocational secondary education Higher education institutions Regional secondary schools, adult education centres, etc. Training vocational centres Language schools and centres Correspondence schools Other educational institutions

Source: Toivonen (2004), pp. 242–243

knowledge-intensive business services, for one reason—the offer of schools is generally directed at individuals, not at companies.[5] As such, it does not meet the business nature of the service.

There are several subdivisions of the KIBS sector. The most popular division, used by many researchers, classifies the KIBS service sector according to their technical nature. This division includes professional KIBS services (P-KIBS) and technical KIBS services (T-KIBS). Although every company offering KIBS uses IT technology to some extent, there are significant differences in the use of this technology to produce KIBS services. For example, market research companies will have to combine the use of IT technology with knowledge of how to conduct research, target groups and how to present and interpret data for the client's needs. In

[5]The exception here are part-time studies, during which the company's employees can acquire knowledge useful in their work.

turn, a company offering computer services will use IT technologies at every stage of creating and delivering its services to clients (Miles, 2012, p. 24).

Despite the too broad definition of the KIBS sector (including the education department), the approach presented by Toivonen (2004) is a good idea that reflects the heterogeneity of the KIBS sector, as it divides the sector into different subcategories (9) to reflect the differences between these types of services. In the case of research on the entire sector, it is worth considering the categories proposed by this author and making analyses at the level of these subcategories. The disadvantage of this approach is the need to have a database based on ISIC Rev.3 classification, which is not always possible. In such a situation, the division into T-KIBS and P-KIBS used in this study seems to be a good solution. As can be seen, defining both the KIBS-type service and the company offering it, and finally the KIBS sector, has been a problem for researchers dealing with these issues for over three decades. The following difficulties can be noted:

- *inconsistency in terminology*—using the term KIBS in the literature to denote both the service and the company or the entire sector;
- *identifying the characteristics of KIBS companies* that are difficult to grasp at the macroeconomic level;
- *evolution of terms*—a different approach of researchers to issues related to knowledge-intensive business services along with the changing conditions of operation of KIBS companies (development of the knowledge-based economy) and new research on the KIBS sector;
- *classification problems*—distinguishing categories/subcategories of KIBS services at different levels of detail in different classifications and assigning different groups/subgroups to the KIBS sector.

In view of the above problems, researching the companies in the KIBS sector is a challenge. First, differences in the definitions and operationalization of the KIBS sector make it difficult to compare studies and create a single integrated picture of the KIBS sector. Secondly, many researchers devote their attention to one type of KIBS sector company, such as IT or consulting companies. Third, although most scholars use the NACE or other identical industrial classification to define the KIBS sector, there are drawbacks to this solution: many new services and activities do not fit into traditional classifications, although they are updated but with little frequency, and at the same time, many companies provide their code of activity at the stage of creating a company, and in the longer term, operate in a different area. Fourth, the KIBS sector is extremely heterogeneous—it includes companies that offer very technical services (e.g. IT companies) and companies that use technologies to a small extent (e.g. law firms), which makes it necessary to group companies in this sector into based on some characteristic features (e.g. the degree of technology use).

There are many reasons why the descriptions of KIBS provided by individual authors differ. First, different researchers have different approaches to issues related to KIBS companies. For example, Miles et al. (1995) do not define the KIBS sector in order to separate it from the service sector, but rather focus on identifying firms with the characteristics of KIBS firms. Secondly, researchers have access to various databases and conduct various types of research on them. Often it is access to data that determines the way in which the KIBS sector is operationalized. An example is the study conducted by Windrum and Tomlinson (1999). Third, researchers base their definitions on various elements—some use, for example, sources of knowledge relevant to KIBS companies, others—industrial classification. The result of such different approaches must be different definitions and classifications of KIBS companies.

Based on the literature analysis in this chapter, the following definitions and approaches are proposed:

1. *Knowledge-Intensive Business Services*—these are services that require the possession and use of the skills to acquire and process specialist knowledge necessary for the functioning of other organizations, providing products/services to their clients and achieving the intended goals, addressed to business entities and created in close cooperation with the client.
2. *Companies offering Knowledge-Intensive Business Services*—companies that combine various types of knowledge and information, both from external and internal sources, to solve a specific problem of their client.
3. *Sector of Knowledge-Intensive Business Services*—sector covering the following types of activity in accordance with the NACE Rev.1.1 classification: 72.1, 72.2, 72.3, 72.4, 72.5, 72.6, 73.1, 73.2, 74.1, 74.2, 74.3, 74.4, 74.5 and the same types of activity in accordance with the NACE Rev.2.0 classification, i.e. divisions 62, 63, 69, 70, 71, 72, 73, 74 and 78 (Schnabl & Zenker, 2013).
4. *Classification of Knowledge-Intensive Business Services*—division into categories P-KIBS (professional KIBS) and T-KIBS (technical KIBS), as proposed in the studies (Koch & Strotmann, 2004), (Koch & Stahlecker, 2006).

The next section presents the specificity of knowledge-intensive business services and companies offering such services.

3.2 Characteristics of Knowledge-Intensive Business Services Companies

This part presents the specifics of knowledge-intensive business services and the companies that offer them. It is difficult to distinguish this specificity between the KIBS service and the KIBS company; therefore, it has been discussed in relation to both concepts together.

3.2.1 Dependency on Knowledge

Knowledge is a component of knowledge-intensive business services, and the dependence of these services on knowledge is an issue often discussed in the literature on the subject and present in many descriptions of KIBS services (Koch & Strotmann, 2005b; Tseng et al., 2011; Tuominen & Toivonen, 2011; Bolisani et al., 2014). In turn, the knowledge possessed by an organization offering KIBS is inextricably linked with the knowledge of its employees. This is especially true for knowledge-intensive business services, where the creation of the service depends almost entirely on the organization's ability to benefit from the knowledge of its employees. Employees employed in the KIBS sector have education in various areas of knowledge: some specialize in the scientific and technical sphere, others—in the administrative, management or socio-legal sphere (Miles, 2005). Apart from higher education in various fields, employees employed in the KIBS sector often have additional features that make them the so-called knowledge workers. These features include, among others creativity, openness to new challenges or awareness of the need for continuous improvement and raising their qualifications, allowing them to create added value in the organization for which they work (Skórska, 2012, p. 106). It is knowledge workers that constitute the main capital of an organization offering knowledge-intensive business services—without their knowledge and the ability to use it, KIBS would not be able to provide their services to clients.

Larsen (2001) developed this approach and proposed the concept of the so-called a distributed knowledge system view, indicating that the way in which knowledge is created is more important than the knowledge possessed by the company and its employees. In other words, the knowledge base is formed by the way of social interaction between KIBS employees and colleagues, colleagues outside the company and customers (creating a so-called community), not the sum of internally available knowledge resources. This has negative consequences for the organization, as it means that the company's knowledge is scattered, it is located in many different places and a single person is not able to know everything. (Larsen, 2001). In summary, dependence on knowledge, or more precisely—on the knowledge possessed by employees of the organization and the ability to create new knowledge as a result of social interactions, is one of the key features of companies offering knowledge-intensive business services.

3.2.2 Adaptation to Customer Needs

The second feature that distinguishes knowledge-intensive business services is the adaptation of these services to the customer's needs, also known as customization. This feature is mentioned in the context of KIBS companies, including in studies (Muller & Doloreux, 2009; Tether & Hipp, 2002). The concept that services are inherently tailored to customer needs is well established in the subject literature as it was introduced almost four decades ago—in 1981—by the Congress of the American Marketing Association. The individualization of services is based on the co-production process (described in the next part of the work). From a certain point of view, the individualization of services is a positive side effect of one of the characteristics of the service, namely the lack of standardization. The inability to produce and offer two identical copies of the service has become an opportunity to better adapt it to the client's needs. Effective interaction between the service provider and his clients leads to the creation or co-production of different versions of the service, perfectly suited to the needs of different clients (Bettiol et al., 2015). Companies offering KIBS services are a good example of this, because many studies to date have emphasized the importance of the relationship between the customer and the KIBS service provider and the highly interactive or customer-dependent nature of this type of service (den Hertog, 2000; Muller & Zenker, 2001a). The strategy opposite to the individualization of the service is its standardization. Many services can be partially or fully standardized through the use of technologies developed for this purpose or through appropriate staff training. Some researchers even take the position that services can be standardized in a manner analogous to manufacturing products—Levitt (1972) proposed an industrialized approach to the production of services, based on production lines. The author even claimed that only standardized services would survive. In practice, this view turned out to be not entirely correct, which is reflected in the growing individualization of services.

The third type of approach, which is a hybrid of the above two, is service modularity. Modularity determines how the system can be broken down into individual components. It helps in solving the problem of the complexity of the service and the resulting problems in designing and creating the service, because it allows the division of tasks when creating a service between individual groups within the organization or outside its borders (Sanchez & Mahoney, 1996). Modularity of services means that the service consists of standard modules containing sub-modules that can be freely mixed and combined with each other. Modularity offers flexible solutions that help customers co-create unique value through the use of multiple service schemes and allow service providers to use their knowledge in customer relationships (Cabigiosu et al., 2015). Bettiol et al. (2015) in their study proposed to distinguish three types of KIBS companies, taking into account the level of individualization of services offered by these companies:

- *KIBS companies offering "tailor-made" services*—companies that offer only fully personalized and tailored services;
- *KIBS companies offering industrialized services*—companies that offer at least one fully standardized service, modular service and standardized service with a limited level of customization;
- *KIBS companies offering dual services*—companies that offer both individualized and standardized services.

Interestingly, from the research conducted by Bettiol et al. (2015), shows that companies offering dual services achieve higher total revenues than companies offering industrialized or "tailor-made" services. Moreover, companies offering dual services employed on average more employees than companies offering the other two types of services. In addition, better results were achieved by companies offering dual services in areas such as market share or the internationalization process. One can risk the statement that while the individualization of KIBS services is an important element in creating and delivering this type of service, it is not the only way for a company from the KIBS sector to achieve good results, because in some cases standardization or modularity of services may also be the right choice.

3.2.3 Co-creation of Value and Co-production of Services

Co-creating value defines how companies engage their customers in the collaborative design and creation of new products, services and innovations (Lehrer et al., 2012). Value is co-created, jointly and mutually, as a result of interactions between service providers and service users, through the integration of resources and the exchange of mutual competences (Vargo et al., 2008). The role of customers in creating service value has been strongly emphasized in studies showing how companies can improve their service offering and/or innovation levels by engaging customers in their own production processes (Cassiman et al., 2009; Gibbert et al. 2002). The client has been perceived as a co-producer of value for some time (Vargo & Lusch, 2004) or its co-creator (Vargo & Lusch, 2008). Moreover, service providers can contribute to customer value by actively exchanging knowledge and participating in the interaction process. While the co-creation of value can only occur when there are mutual interactions between the firm and the client, the mere existence of an interaction does not automatically mean that the firm is committed to co-creating value with the client. The company's activities in the form of direct interaction with the client can have both a positive (value creation) and a negative impact (value destruction) on the relationship with the client (Gronroos, 2011). Recognizing as value creation the process by which the client's condition or well-being is improved (Vargo et al., 2008), it can be stated that as a result of inadequate conduct of this process, the client's well-being or condition may deteriorate, which may result in the client's reluctance to use the services of a given KIBS company.

Another feature, related in some way to the previous one, is co-production, also known as co-creation,[6] of KIBS. Co-production is the degree to which the customer is involved in producing and delivering the service (Ordanini & Pasini, 2008). The cooperation between KIBS companies and their clients depends on many factors, such as: a clear way of formulating the client's problem, the possibility of solving it by KIBS, especially in terms of human resources, and the motivation of employees to perform the task (Skórska, 2012, p. 111). Co-production is usually a feature of business services because providing them requires the company offering them to make their business resources available, not just time, attention and personal ability, as is the case with consumer services (Gallouj & Savona, 2009). The client's participation in the development of the KIBS service reflects his active role in relation to activities, inputs or behaviours, not just mere presence at the time of service provision. The outlays may be physical (e.g. material property of the client, his physical effort), emotional (e.g. patience, empathy) or intellectual

[6]The concept of co-production is also connected with the idea of prosumption, i.e. the client fulfilling two roles simultaneously: producer and consumer (Toffler, 1980).

(e.g. information obtained and provided by the client to KIBS) (Skórska, 2012, p. 111).

There are a number of factors that affect the effectiveness of KIBS co-production. Customers must be adequately motivated to participate in the co-production and must understand the impact of their complicity on the final form of the service. In addition, at various stages of creating this type of service, it is necessary to exchange knowledge and information between the client and the KIBS company. Bettencourt et al. (2002) identified seven characteristics of effective co-production of the KIBS service: open communication, joint problem solving, tolerance, compromise, accommodating, support, commitment to project supervision and personal dedication.

Close relationships between service providers and customers require coordination of activities. This is especially important when the nature of the relationship between KIBS and the client, as well as the end result of their cooperation, is complex and tailored to the client's needs. Developing an optimal knowledge-based solution that meets the project's and client's goals requires open communication and the client shares all information that may be relevant to the project. This type of communication often requires the client to share in-depth knowledge of business processes, strategic goals and the industry and market conditions in which they compete. Importantly, the client must also openly communicate project information and relationship expectations as well as their project vision and priorities. These types of communication behaviours are especially important in the early stages of a project when analyzing the nature of the problem and formulating possible solutions. Optimal knowledge-intensive solutions require mutual commitment and cooperation from both the client and KIBS. The client contributes his approach and unique sources of competences that must be used to create the KIBS service. Moreover, it is necessary to solve problems together, because complex solutions often require various types of concessions and adjustments. These kinds of situations require clients to be critical and receive critical active rather than passive role in creating solutions. This involves taking the initiative to communicate potential problems or gaps in the solutions being developed, ask thoughtful questions during meetings and provide constructive feedback on proposed solutions.

Constant uncertainty and joint coordination of the creation of the KIBS service irrevocably lead to complications or difficulties. In such a situation, it is important for the client to show tolerance and understanding that will support the building of the relationship rather than creating mounting tension. A customer who reacts tolerantly to minor disruptions in the creation of the KIBS service will contribute to strengthening the relationship with the supplier and a functional solution to the problem. Compromise means the level to which the client is able to accept KIBS ideas and suggestions. The determinants of effective cooperation include flexibility and the tendency to adopt the views of the other party. Lack of trust and faith in the skills and opinions of the service provider makes it difficult to find the optimal solution to the problem.

Accommodation is about mutual inclusion or taking into account the needs and approaches of each party (both KIBS and its clients). Optimal KIBS service must reflect the perceptions of both the customer and the service provider. The success of the KIBS service often depends on the active involvement and participation of many actors in the client's company, including the end users of the KIBS service. Taking into account the fact that successful projects require internal support in the client's company, it is also important to properly "sell" the project to the members of the organization and their involvement in the implementation of the adopted solutions. People responsible for co-creating the KIBS service should be able to create an atmosphere of acceptance and unity around the developed solutions, presenting their advantages and benefits. Involvement in supervision over the project means the client's participation in controlling the implementation of the KIBS service in relation to the costs and time of this implementation. Clients can act as an additional element of control to the service provider, ensuring that everything is moving towards the project goals and is going as expected. Clients who understand the importance of co-production in knowledge-intensive services are actively involved in project management at all times to ensure that the project achieves its goals by monitoring progress and service provider, keeping abreast of key project issues and acting as a liaison with staff inside the company. The client can request reports on the progress of the project implementation or participate in working meetings. The last essential element is the personal dedication of a key person on the client's side. Although many people in the client's company may be involved in the work related to the KIBS service, the personal involvement of a key employee is especially important for effective partnership and achievement of goals. This person acts as the liaison between KIBS and the customer organization and makes decisions on behalf of the customer (Bettencourt et al., 2002).

As research shows, the optimal level of co-production varies between different types of KIBS companies (Lehrer et al., 2012). While in IT companies the level of co-production is and usually has to be high, in the case of companies in the design sector (e.g. engineering, design services) the situation is different. The co-production of the KIBS service will also be different for different types of these services. There are four types of roles that customers can play in co-authoring services (Kuusisto, 2008). The first is the role of the consumer. In this case, the customer is only a user of the service provided by KIBS. The client is usually aware of his needs and expectations towards the service provider and controls the specification of the service. Service specificities can be clarified by either party, and feedback is usually collected after the service has been performed. An example is the standard legal advice service. The second role is the role of the co-performer (co-performer role). In this role, the client performs the tasks necessary to provide the service. The client serves as a "component" of the production of services in the sense that it carries out tasks as planned or designed by the service provider (e.g. developing an IT application for the client). The third possible role is a co-creator role in which the customer co-creates the solution with the service provider. In this case, the service is essentially created as a result of the interaction between the customer and the service provider. Here, the client is actually a contributor to the solution to the problem/situation to be found. An example would be a strategic research and development advisory service. The fourth and final role is the role of a co-designer or partner (co-designer role). In this case, the client's influence on the creation of the service is the greatest and together with the service provider he decides about the division of the workforce and the method of cooperation.

The issue of co-creating value and co-producing KIBS services in a more practical dimension was analysed by Lessard (2015). In her article, she defined the requirements for modelling the relationship of KIBS companies with their clients in a way that takes into account the specific features of the KIBS service and can support its co-creation. The author also proposed a technique of Value Co-creation Modelling (VCM), which allows you to measure and evaluate the elements that contribute to the creation of value for individual stakeholders at the level of the process, result and the possibility of achieving it.

3.2.4 Creating and Supporting Innovation

KIBS companies are often examined in the context of participating in the innovation processes of their clients and producing innovations for their own use. As stated by Doloreux and Shearmur (2010), KIBS companies can play the role of "streamliners" of innovation by supporting their customers in creating them or transferring knowledge generated elsewhere to the customer's company, and they can also act as innovators by introducing new (possibly improved) services, production methods or organizational solutions. In the first case, they act as intermediaries between clients and external sources of knowledge (den Hertog, 2000), in the second, they undertake innovative activities in areas such as technological, marketing and organizational solutions (Evangelista, 2006). KIBS companies can be assigned three functions in the economy:

- *Facilitators of innovation*—a company providing business knowledge-intensive services supports the company—the client in its innovation processes, but is not the creator of innovation;
- *Carriers of innovation*—the company offering KIBS plays a role in transferring an existing, innovative solution from company to company—customer. In this case, as in the above, the innovation itself is not the result of the work of the company offering KIBS;
- *Sources of innovation*—in this case, the company offering business knowledge-intensive services plays a key role in initiating and developing innovation in the client's company (Hauknes & Antonelli, 1999).

The second of the mentioned roles—the carrier of innovation—is also sometimes referred to as an innovation broker or a "bridge" for innovation (Czarnitzki & Spielkamp, 2003). Other roles for companies from the KIBS sector were distinguished by E. Muller, according to whom KIBS companies perform the following functions for their companies—clients from the small- and medium-sized enterprise (SME) sector:

- performance of an analysis of the external environment relating to innovative activities of SMEs. This analysis does not necessarily have to be limited to technical aspects, it can also be related to areas such as finance, marketing or law;
- carrying out an internal analysis aimed at strengthening the innovative capacity of SMEs;
- co-production of innovation through: (1) support in the field of applied methods, (2) support related to organizational change, (3) direct participation in the creation of the concept and development of innovation;
- accompanying SMEs during the phase of introducing innovations to the market (Muller, 2001, p. 40).

The influence of KIBS companies on the innovativeness of their companies—clients has been confirmed in many studies, e.g. (Aslesen & Isaksen, 2007b; Toivonen, 2004; Yam et al., 2011).

KIBS companies behave differently in relation to their own innovations. Rodriguez and Camacho (2010) have defined four such roles: "hard" innovators, "soft" innovators, lone innovators and knowledge diffusers. The first category is companies whose innovative activity is similar to modern production plants, the second category is those that invest in innovation expenditure other than R&D activity, the third category are companies that develop innovations on their own and the fourth category are companies with high the degree of interaction with local agents of the innovation system. The effect of these behaviours may be various types of innovations:

- *product innovations*—relate to the placing on the market of new or improved products or services;
- *process innovations*—relate to the introduction of new or significantly improved processes (e.g. production but not related to delivery, unless it is an integral part of the production process);
- *delivery innovations*—refer to changes in the way an entity delivers its products or services to its customers, e.g. introducing just-in-time delivery, introducing new or significantly improved home delivery services;
- *strategic innovations*—relate to the implementation of new or significantly modified business strategies, e.g. targeting different markets, implementing new or significantly modified corporate missions;
- *management innovations*—relate to the implementation of new or significantly modified managerial techniques, e.g. introducing knowledge management or quality management practices;
- *marketing innovations*—relate to the implementation of new or significantly modified marketing strategies and concepts, e.g. the introduction of new or significantly improved marketing methods (Amara et al., 2009).

In a study conducted by Zięba (2015) on a sample of 111 enterprises from the KIBS sector, it was shown that the surveyed companies more often declared the introduction of incremental product/service innovations than radical ones[7] (the former were introduced by 96 companies, while the latter—69; some companies have introduced two types of innovation). Similarly differentiated results concerned incremental and radical process innovations—incremental innovations were introduced by 97 companies, while radical innovations were introduced by 68; here too, some companies have introduced two types of innovation (Zięba, 2015).

To sum up, KIBS companies play a very important role in the economy, because, on the one hand, they generate various kinds of innovations themselves, and on the other hand—they contribute to creating innovations in their clients' companies. This is very important especially for companies from the SME sector, which face many obstacles on a daily basis, taking actions aimed at creating innovations.[8] Overall, KIBS companies are seen as innovative and supportive of innovation.

3.2.5 Regional Concentration and Influence on the region's Economy

Another feature of the KIBS sector companies is their regional concentration. It has been empirically proven that KIBS companies are decidedly concentrated in the fastest-growing urban areas, and a significant increase in their number in urban regions across Europe in just a few decades is positively correlated with overall economic growth (Strambach, 2010, p. 176). In other words, companies from the KIBS sector tend to clustering in metropolitan areas (Wood et al., 1993), in peripheral regions, however, their share is limited. Strong tendencies to accumulate were confirmed, inter alia, by in a study of KIBS companies located in the Milan metropolitan region (Antonietti et al., 2013). This phenomenon is conditioned by various reasons, which researchers unanimously classify as demand and supply factors (Aslesen & Isaksen, 2007b). The former include, for example, proximity to customers, the ability to reach a large number of recipients or establish cooperation with other entities, while the latter include factors such as, e.g., developed infrastructure or access to resources, e.g. skilled workers.

According to Skórska (2012, p. 115), the genesis of this concentration can be traced back to the 1970s, when the geographical development of production services depended on urbanized regions and large agglomerations in which there was a concentration of industrial activity, and thus—the demand for production services was also high. In such regions, there was easier access to sources of knowledge (e.g. universities or research institutes), various types of resources (e.g. skilled labour) or outlets. It was also possible to facilitate cooperation with the client of

[7]Incremental innovations are improvements and improvements in the existing products, services and processes, and radical innovations consist in the introduction of completely new, previously non-existent products, services and processes (Bhaskaran, 2006).

[8]Examples of obstacles faced by companies from the SME sector may include: lack of capital, lack of management skills, difficulties in obtaining technological knowledge and the so-called the know-how needed to create innovation (Muller & Zenker, 2001a, p. 1505).

the service, if only due to the proximity of the location. The importance of the latter factor is confirmed by the research conducted by Keeble and Nachum (2002). These researchers analysed the national and regional characteristics of KIBS companies in the UK, including a detailed study of 300 small- and medium-sized KIBS companies located in central London, East England and the South West of England, and found that the concentration of KIBS companies results, inter alia, from the need and benefit of proximity and availability of customers. Similar conclusions were drawn by Koch and Stahlecker, who examined the relationship between the creation of KIBS companies and their innovation in the German metropolitan regions of Bremen, Munich and Stuttgart (Koch & Stahlecker, 2006). According to the researchers, the geographical proximity of suppliers and customers is crucial—especially in the early stages of establishing a new KIBS company. Moreover, their analyses showed that the structure and configuration of the regional knowledge base can play an important role in the development of these start-ups. The results of the study carried out in Romania also show that the activities of KIBS companies are characterized by a significant spatial concentration in urban centres, especially in Bucharest. The spatial concentration of such companies was also clear in several Romanian counties: Braşov, Cluj, Constanţa, Ilfov, Sibiu and Timiş (Bujdosó et al., 2016). As KIBS companies are also buyers of services and goods, location in metropolitan/urbanized areas brings them additional benefits, which include: local availability of specialized suppliers, greater likelihood of face-to-face interaction, as well as lower transport and transaction costs with due to increased mutual visibility in the market and a higher level of mutual trust (Helsley & Strange, 2007).

Due to the concentration of KIBS companies in some regions, they are seen as a driving force for the development of local economies (Smedlund & Toivonen, 2007; Wood, 2006). On the one hand, KIBS companies help in development, and on the other—their formation depends on the economic conditions of a given region and the existing innovation system (Huggins, 2011). The presence of KIBS companies contributes to the growth of investments in the region, which increases the demand for KIBS services and, consequently, also their greater supply. In the peripheral regions, the lower popularity of KIBS services is severely felt by companies from the SME sector, which are often characterized by a lack of professional knowledge. In such regions, the problem is also the lower benefits derived from the diffusion of KIBS innovative activities compared to urbanized regions. A solution to the problem of differences in development may be an appropriate regional and national policy, supporting the development of KIBS companies also in less urbanized areas (Skórska, 2012, p. 115).

3.2.6 Networking

Due to the specificity of KIBS companies and their dependence on other entities in acquiring or creating knowledge, another characteristic feature of companies in this sector is their network activity. Companies from the KIBS sector often interact not only with their customers, but also suppliers and even competitors, and together with them they create a service and added value for all entities. Networks are increasingly the subject of research, as thanks to the inter-organizational knowledge flows they contribute to the development of innovation (Capaldo, 2007; Kogut, 2000). Czakon and Klimas underline the importance of innovation networks[9] for knowledge-intensive industries, even suggesting that they shape the conditions for running a business by the organizations that contribute to these networks (Czakon & Klimas, 2014). Moreover, the owned social network determines the companies' access to information, thus diversifying their possibilities of obtaining relevant information and knowledge, and thus—access to potential sources of competitive advantage.

In the interaction sector of KIBS, personal networks can be very important to the growth of a business as knowledge, learning and interaction play a unique role in service creation in general, and in the case of non-standardized services that role is even greater. A project-oriented way of delivering specialist services requires constant learning and networking—therefore personal networks play a very important role in the KIBS sector. In this case, personal networks refer not only to the person of the owner or manager of this type of company, but also to rank and file employees who often closely cooperate with clients and intensively acquire knowledge from various market actors. In the KIBS sector, functional links between companies and their customers, knowledge providers and partners are also of particular importance (Koch & Strotmann, 2004).

In the case of small KIBS companies, a very important role is played by the network relations of the owner who, by increasing his own network contacts, can create new patterns of action in cooperation with knowledge-intensive people and organizations supporting a given company and its internal development (Siikonen et al., 2011). Moreover, for the development of a small KIBS company, organizations at the regional and national level are important, and contacts with educational and research institutions are particularly important. Network relationships play a vital role in KIBS strategies for products/services, production processes, markets, innovation, etc., as well as in their success. Therefore, as the aforementioned researchers suggest, for the development of the KIBS sector to be possible, measures should be taken to develop network skills and opportunities among small KIBS companies.

Companies from the KIBS sector can use one of the two networking strategy approaches. The first is an open approach and the second is a closed approach (Paiola et al., 2013).

- *Open networking strategy*—refers to companies that frequently and intensively engage in interactions with external partners that form the basis of their business. An open network approach means that the company wants to share knowledge

[9]Czakon and Klimas (2014), following (Dhanaraj & Parkhe, 2006), define an innovative network as a set of organizations and relations between them, oriented towards acquiring information, access to and sharing knowledge, continuous learning, R&D and commercialization of knowledge.

with partners (suppliers, customers, service providers, etc.) in order to use its competences and acquire additional skills developed elsewhere. This approach usually characterizes companies that cannot manage all the knowledge required to run a business or do not want to bear the high risk associated with developing this knowledge in the company.
- *Closed networking strategy*—refers to companies that develop and explore their internal competences and do not have to or do not want to share elements of their knowledge with external partners. At the same time, these companies are reluctant to use external sources of knowledge and do not interact in the network with partners.

In the case of an open network strategy, two successive approaches can be distinguished, according to the "direction" (up or down) of the network relations (Paiola et al., 2013). These approaches can be defined as follows:

- *Customer approach*—refers to companies that interact and share knowledge, especially with customers. In other words, it is primarily from customers that the company obtains useful elements of knowledge that enable its development.
- *Approach towards suppliers*—it is symmetrical to the previous one and refers to companies that share their knowledge mainly with suppliers. In other words, the basis of KIBS business is the knowledge gained through interaction with suppliers (in particular with technology suppliers).

It is proposed to extend the above typology to include the following categories of network strategy:

- *Network strategy towards competitors*—refers to companies that interact and share knowledge with their competitors. As a result of this interaction, they acquire useful elements of knowledge for development and supplement their competences. This surprising strategy may be useful especially in the case of small- and medium-sized KIBS companies, which do not have the appropriate and sufficient resources to offer their services on a large scale, and thanks to cooperation with competitors can not only meet the needs of their clients, but also acquire new knowledge.
- *Network strategy towards research and development institutes*—refers to companies that work closely with various kinds of scientific and research institutions, e.g. universities, R&D centres, etc. This kind of strategy may be necessary for T-KIBS companies that rely heavily on technological development and they must supplement their knowledge with current technologies and tools.

Summing up, it can be stated that KIBS companies often enter into various types of network interactions with various entities operating on the market, such as customers, suppliers, distributors or research and development institutions. Although the network strategies selected by KIBS companies may differ (e.g. open/closed, customer or supplier oriented), they are an important element in the operation of these companies as they help them acquire the knowledge needed to carry out their own activities.

3.3 Summary

This chapter presented the characteristics of the KIBS sector and the companies operating in it. An overview of the approaches to the KIBS sector, companies offering KIBS services as well as this type of services was presented. Based on the analysed literature, a definition of KIBS services and other basic terms were also proposed. Another issue discussed in the chapter was the sector of knowledge-intensive business services along with the types of activities included in this sector. An overview of the classifications of this sector proposed by various researchers was also presented.

After presenting the issues related to the definition and classification of the KIBS sector, the specificity of knowledge-intensive business services and the companies offering them were illustrated. It can be seen that these services are indeed very specific and require more attention from the academic world.

References

Amara, N., Landry, R., & Doloreux, D. (2009). Patterns of innovation in knowledge-intensive business services. *The Service Industries Journal, 29*(4), 407–430. https://doi.org/10.1080/02642060802307847

Amara, N., Landry, R., & Traoré, N. (2008). Managing the protection of innovations in knowledge-intensive business services. *Research Policy, 37*(9), 1530–1547. https://doi.org/10.1016/j.respol.2008.07.001

Antonietti, R., Cainelli, G., & Lupi, C. (2013). Vertical disintegration and spatial co-localization: The case of KIBS in the metropolitan region of Milan. *Economics Letters, 118*(2), 360–363. https://doi.org/10.1016/j.econlet.2012.11.031

Aslesen, H. W., & Isaksen, A. (2007a). Knowledge intensive business services and urban industrial development. *The Service Industries Journal, 27*(3), 321–338.

Aslesen, H. W., & Isaksen, A. (2007b). New perspectives on knowledge-intensive services and innovation. *Geografiska Annaler, 89B*(S1), 45–58.

Baláž, V. (2004a). Knowledge-intensive business services in transition economies. *The Service Industries Journal, 24*(4), 83–100. https://doi.org/10.1080/0264206042000275208

Baláž, V. (2004b). Patterns of intermediate consumption and productivity in the knowledge intensive business services in transition economies. *Ekonomicky Casopis, 52*(3), 298–314.

Bettencourt, L. A., Ostrom, A. L., Brown, S. W., & Roundtree, R. I. (2002). Client co-production in knowledge-intensive business services. *California Management Review, 44*(4), 100–128. https://doi.org/10.2307/41166145

Bettiol, M., De Marchi, V., Di Maria, E., & Grandinetti, R. (2012). Determinants of market extension in knowledge-intensive business services: Evidence from a regional innovation system. *European Planning Studies, 21*(4), 1–18. https://doi.org/10.1080/09654313.2012.722930

Bettiol, M., Di Maria, E., & Grandinetti, R. (2015). Service customisation and standardisation in combinatory knowledge-intensive business services. *International Journal of Knowledge-Based Development, 6*(3), 241. https://doi.org/10.1504/IJKBD.2015.072812

Bhaskaran, S. (2006). Incremental innovation and business performance: Small and medium-size food enterprises in a concentrated industry environment. *Journal of Small Business Management, 44*(1), 64–80. https://doi.org/10.1111/j.1540-627X.2006.00154.x

Bolisani, E., Paiola, M., & Scarso, E. (2014). Cognitive features of KIBS companies: Evidence from a cluster analysis. *International Journal of Services Technology and Management, 20* (4–6), 215–232. https://doi.org/10.1504/IJSTM.2014.068853

Bujdosó, Z., Pénzes, J., Dávid, L., & Madaras, S. (2016). The spatial pattern of KIBS and their relations with the territorial development in Romania. *Amfiteatru Economic, 18*(41), 73–88.

Büttgen, M., & Ates, Z. (2009). Customer participation and its effects on service organisations: An institutional economics perspective. In *Proceedings of the 2009 Naples Forum on Service*, 1–45. http://www.naplesforumonservice.it/uploads//files/ATES_CUSTOMER PARTICIPATION AND ITS EFFECTS ON SERVICE ORGANISATIONS_AN INSTITUTIONAL ECONOM ICS PERSPECTIVE(1).pdf.

Cabigiosu, A., Campagnolo, D., Furlan, A., & Costa, G. (2015). Modularity in KIBS: The case of third-party logistics service providers. *Industry and Innovation, 22*(2), 126–146. https://doi.org/10.1080/13662716.2015.1023012

Capaldo, A. (2007). Network structure and innovation: The leveraging of a dual network as a distinctive relational capability. *Strategic Management Journal, 28*, 585–608. https://doi.org/10.1002/smj

Cassiman, B., Di Guardo, M. C., & Valentini, G. (2009). Organising R&D projects to profit from innovation: Insights from co-opetition. *Long Range Planning, 42*(2), 216–233. https://doi.org/10.1016/j.lrp.2009.01.001

Ciriaci, D., Montresor, S., & Palma, D. (2015). Do KIBS make manufacturing more innovative? An empirical investigation of four European countries. *Technological Forecasting and Social Change, 95*, 135–151. https://doi.org/10.1016/j.techfore.2015.02.008

Ciriaci, D., & Palma, D. (2016). Structural change and blurred sectoral boundaries: Assessing the extent to which knowledge-intensive business services satisfy manufacturing final demand in Western countries. *Economic Systems Research, 28*(1), 55–77. https://doi.org/10.1080/09535314.2015.1101370

Consoli, D., & Elche-Hortelano, D. (2010). Variety in the knowledge base of knowledge intensive business services. *Research Policy, 39*(10), 1303–1310. https://doi.org/10.1016/j.respol.2010.08.005

Czakon, W., & Klimas, P. (2014). Innovative networks in knowledge-intensive industries – How to make them work? An empirical investigation into the polish Aviation Valley. In D. Jemielniak (Ed.), *The laws of the knowledge workplace. Changing roles and the meaning of work in knowledge-intensive environments* (pp. 133–157). Gower Publishing Limited.

Czarnitzki, D., & Spielkamp, A. (2003). Business services in Germany: Bridges for innovation. *The Service Industries Journal, 23*(2), 1–30. https://doi.org/10.1080/02642060412331300862

den Hertog, P. (2000). Knowledge-intensive business services as co-producers of innovation. *International Journal of Innovation Management, 4*(4), 491–528.

Desmarchelier, B., Djellal, F., & Gallouj, F. (2013). Knowledge intensive business services and long term growth. *Structural Change and Economic Dynamics, 25*(1), 188–205. https://doi.org/10.1016/j.strueco.2012.07.003

Deza, X. V., & Lopez, M. G. (2014). Regional concentration of knowledge-intensive business services in Europe. *Environment and Planning C: Government and Policy, 32*(6), 1036–1058. https://doi.org/10.1068/c11171r

Dhanaraj, C., & Parkhe, A. (2006). Orchestrating innovation networks. *Academy of Management Review, 31*(3), 659–669. https://doi.org/10.1007/978-3-540-92267-4

Doloreux, D., & Shearmur, R. (2010). Exploring and comparing innovation patterns across different knowledge intensive business services. *Economics of Innovation and New Technology, 19*, 605–625. https://doi.org/10.1080/10438590903128966

Doroshenko, M., Miles, I., & Vinogradov, D. (2013). Knowledge intensive business services as generators of innovations. *Higher School of Economics Research Paper No. WP BRP 12/STI/ 2013, May*. https://doi.org/10.2139/ssrn.2282511

Doroshenko, M. E. (2012). How knowledge-intensive business services upgrade their customers: Evidence from Russia. In E. Di Maria, R. Grandinetti, & B. Di Bernardo (Eds.), *Exploring knowledge-intensive business services: Knowledge management strategies* (pp. 79–99). Palgrave Macmillan UK. https://doi.org/10.1057/9781137008428_5

European Commission. (2009). Challenges for EU support to innovation in services. Fostering new markets and jobs through innovation. https://doi.org/10.2769/11963

Evangelista, R. (2006). Innovation in the European service industries. *Science and Public Policy, 33* (9), 653–668. https://doi.org/10.3152/147154306781778632

Fernandes, C., Ferreira, J. J., & Marques, C. S. (2013). Innovation management capabilities in rural and urban knowledge intensive business services: Empirical evidence. *Service Business*, 1–24. https://doi.org/10.1007/s11628-013-0225-7

Ferreira, J. J. M., Raposo, M. L., & Fernandes, C. I. (2013). Does innovativeness of knowledge-intensive business services differ from other industries? *The Service Industries Journal, 33* (7–8), 734–748. https://doi.org/10.1080/02642069.2013.740462

Freel, M. (2006). Patterns of technological innovation in knowledge-intensive business services. *Industry & Innovation, 13*(3), 335–358. https://doi.org/10.1080/13662710600859157

Gadrey, J., & Gallouj, F. (2002). *Productivity, innovation and knowledge in services: New economic and socio-economic approaches*. Edward Elgar.

Gallouj, F., & Savona, M. (2009). Innovation in services: A review of the debate and a research agenda. *Journal of Evolutionary Economics, 19*(2), 149–172. https://doi.org/10.1007/s00191-008-0126-4

Gibbert, M., Leibold, M., & Probst, G. (2002). Five styles of customer knowledge management, and how smart companies use them to create value. *European Management Journal, 20*(5), 459–469. https://doi.org/10.1016/S0263-2373(02)00101-9

Gronroos, C. (2011). Value co-creation in service logic: A critical analysis. *Marketing Theory, 11* (3), 279–301. https://doi.org/10.1177/1470593111408177

Hauknes, J., & Antonelli, C. (1999). Knowledge intensive services – What is their role? In *Paper for the OECD Business and Industry Policy Forum, STEP Group* (issue September). STEP Group.

Helsley, R. W., & Strange, W. C. (2007). Agglomeration, opportunism, and the organization of production. *Journal of Urban Economics, 62*(1), 55–75. https://doi.org/10.1016/j.jue.2006.05.006

Herstad, S. J., & Ebersberger, B. (2014). Urban agglomerations, knowledge-intensive services and innovation: Establishing the core connections. *Entrepreneurship & Regional Development, 26* (3–4), 211–233. https://doi.org/10.1080/08985626.2014.888098

Hipp, C. (1999). Knowledge-intensive business services in the new mode of knowledge production. *AI & SOCIETY, 13*(1–2), 88–106. https://doi.org/10.1007/BF01205260

Huggins, R. (2011). The growth of knowledge-intensive business services: Innovation, markets and networks. *European Planning Studies, 19*(8), 1459–1480. https://doi.org/10.1080/09654313.2011.586172

Keeble, D., & Nachum, L. (2002). Why do business service firms cluster? Small consultancies, clustering and decentralization in London and southern England. *Transactions of the Institute of British Geographers, 27*(1), 67–90. https://doi.org/10.1111/1475-5661.00042

Koch, A., & Strotmann, H. (2004). The impact of regional and functional integration on the post-entry performance of knowledge intensive business service firms. In *44th Congress of the European Regional Science Association*.

Koch, A., & Strotmann, H. (2005a). *The impact of functional integration and spatial proximity on the post-entry performance of knowledge intensive business service firms* (No. 18; IAW-Diskussionspapiere).

Koch, A., & Strotmann, H. (2005b). Determinants of innovative activity in newly founded KIBS. In *3rd Interdisciplinary European Conference on Entrepreneurship Research (IECER)*, 1–41.

Koch, A., & Stahlecker, T. (2006). Regional innovation systems and the foundation of knowledge intensive business services. A comparative study regional innovation systems and the foundation of knowledge intensive business services. A comparative study in Bremen, Munich, and Stuttgart. *European Planning Studies, 14*, 37–41.

Koch, A., & Strotmann, H. (2008). Absorptive capacity and innovation in the knowledge intensive business service sector. *Economics of Innovation and New Technology, 17*(6), 511–531. https://doi.org/10.1080/10438590701222987

Kogut, B. (2000). The network as knowledge: Generative rules and the emergence of structure. *Strategic Management Journal, 21*(3), 405. https://doi.org/10.2307/3094194

Koschatzky, K., & Stahlecker, T. (2006). Structural couplings of young knowledge-intensive business service firms in a public-driven regional innovation system. *Entrepreneurship in the Region, 14*, 171–193. https://doi.org/10.1007/0-387-28376-5

Kuusisto, A. (2008). Customer roles in business service production – Implications for involving the customer in service innovation. In *TUTKIMUSRAPORTTI – RESEARCH REPORT* (Vol. 195). http://eprints.herce.fi/126/

Larsen, J. N. (2001). Knowledge, human resources and social practice: The knowledge-intensive business service firm as a distributed knowledge system. *Service Industries Journal, 21*(1), 81–102. https://doi.org/10.1080/714004998

Lehrer, M., Ordanini, A., DeFillippi, R., & Miozzo, M. (2012). Challenging the orthodoxy of value co-creation theory: A contingent view of co-production in design-intensive business services. *European Management Journal, 30*(6), 499–509. https://doi.org/10.1016/j.emj.2012.07.006

Lessard, L. (2015). Modeling value Cocreation processes and outcomes in knowledge-intensive business services engagements. *Service Science, 7*(3), 181–195. https://doi.org/10.1287/serv.2015.0104

Levitt, T. (1972). Production-line approach to service. *Harvard Business Review, 50*(5), 41–52.

Miles, I. (2005). Knowledge intensive business services: Prospects and policies. *Foresight, 7*(6), 39–63. https://doi.org/10.1108/14636680510630939

Miles, I. (2008). *Knowledge-intensive services in Europe*. Manchester Institute for Innovation Research, Manchester Business School, University of Manchester, Manchester, September. www.academia.edu/235223/Knowledge_Intensive_Services_in_Europe

Miles, I. (2011). From knowledge-intensive services to knowledge-intensive service systems BT – Special issue on knowledge intensive service systems. *International Journal of Services, Technology and Management, 16*(2), 141–159. https://doi.org/10.1504/IJSTM.2011.042593

Miles, I. (2012). KIBS and knowledge dynamics in client-supplier interaction. In E. Di Maria, R. Grandinetti, & B. Di Bernardo (Eds.), *Exploring knowledge-intensive business services: Knowledge management strategies* (pp. 13–34). Palgrave.

Miles, I., Kastrinos, N., Flanagan, K., Bilderbeek, R., Den Hertog, P., Huntink, W., & Bouman, M. (1995). Knowledge-intensive business services. Users, carriers and sources of innovation. *A Report to DG13 SPRINT-EIMS, March*, 1–117.

Miozzo, M., & Grimshaw, D. (Eds.). (2006). *Knowledge intensive business services. Organizational forms and national institutions*. Edward Elgar. https://doi.org/10.4337/9781847201751

Muller, E. (2001). *Innovation interactions between knowledge-intensive business services and small and medium-sized enterprises*. Physica-Verlag. A Springer-Verlag Company.

Muller, E., & Doloreux, D. (2009). What we should know about knowledge-intensive business services. *Technology in Society, 31*(1), 64–72. https://doi.org/10.1016/j.techsoc.2008.10.001

Muller, E., & Zenker, A. (2001a). Business services as actors of knowledge transformation: The role of KIBS in regional and national innovation systems. *Research Policy, 30*(9), 1501–1516. https://doi.org/10.1016/S0048-7333(01)00164-0

Muller, E., & Zenker, A. (2001b). Business services as actors of knowledge transformation and diffusion: Some empirical findings on the role of KIBS in regional and national innovation systems. In *Working Papers Firms and Region* (R2/2001; Working Papers Firms and Region).

Nählinder, J. (2005). Innovation and employment in services: The case of knowledge intensive business Services in Sweden (Linköping).

Nählinder, J., & Hommen, L. (2002). Employment and innovation in services: Knowledge intensive business services in Sweden. ...: *Linkoping University, Department of Technology and ..., April*, 1–25. http://scholar.google.com/scholar?hl=en&btnG=Search&q=intitle:Employment+and+Innovation+in+Services+:+Knowledge+Intensive+Business+Services+in+Sweden#0

O'Farrell, P. N., & Moffat, L. A. R. (1995). Business services and their impact upon client performance: An exploratory interregional analysis. *Regional Studies, 29*, 111–124. https://doi.org/10.1080/00343409512331348843

Ordanini, A., & Pasini, P. (2008). Service co-production and value co-creation: The case for a service-oriented architecture (SOA). *European Management Journal, 26*(5), 289–297. https://doi.org/10.1016/j.emj.2008.04.005

Paiola, M., Bolisani, E., & Scarso, E. (2013). Characterisation of knowledge-based networking strategies: Insights from the KIBS sector. *International Journal of Networking and Virtual Organisations, 13*(3), 263–281.

Pardos, E., Gómez-Loscos, A., & Rubiera-Morollón, F. (2007). 'Do versus buy' decisions in the demand for knowledge intensive business services. *The Service Industries Journal, 27*(3), 233–249. https://doi.org/10.1080/02642060701206991

Ritala, P., Hyötylä, M., Blomqvist, K., & Kosonen, M. (2013). Key capabilities in knowledge-intensive service business. *The Service Industries Journal, 33*(5), 486–500. https://doi.org/10.1080/02642069.2011.623774

Rodriguez, M., & Camacho, J. A. (2010). Are knowledge-intensive business services so "hard" innovators? Some insights using Spanish microdata. *Journal of Innovation Economics and Management, 1*(5), 41–65. https://doi.org/10.3917/jie.005.0041

Sanchez, R., & Mahoney, J. T. (1996). Modularity, flexibility, and knowledge management in product and organization design. *Strategic Management Journal, 17*(S2), 63–76. https://doi.org/10.1002/smj.4250171107

Scarso, E., & Bolisani, E. (2012). Trust in knowledge exchanges between service providers and clients: A multiple case study of KIBS. *Knowledge Management Research & Practice, 10*(1), 16–26. https://doi.org/10.1057/kmrp.2011.28

Schnabl, E., & Zenker, A. (2013). Statistical classification of knowledge-intensive business services (KIBS) with NACE rev. 2. *EvoREG Research Note, 25* (June), 1–11. papers3://publication/uuid/7AB446C4-BC91-44AF-8EB2-A5BC5950EB04.

Shi, X., Wu, Y., & Zhao, D. (2014). Knowledge intensive business services and their impact on innovation in China. *Service Business, 8*(4), 479–498. https://doi.org/10.1007/s11628-013-0202-1

Siikonen, J., Pellikka, J., & Littunen, H. (2011). Knowledge networks in innovation among small KIBS firms. *International Journal of Innovation and Regional Development, 3*(5), 515–534.

Simmie, J., & Strambach, S. (2006). The contribution of KIBS to innovation in cities: An evolutionary and institutional perspective. *Journal of Knowledge Management, 10*(5), 26–40. https://doi.org/10.1108/13673270610691152

Skogli, E. (1998). Knowledge intensive business services: A second national knowledge infrastructure? In *STEP Working Paper: Vol. A-03*.

Skórska, A. (2012). *Wiedzochłonne usługi biznesowe w Polsce i innych krajach Unii Europejskiej*. Wydawnictwo Uniwersytetu Ekonomicznego w Katowicach.

Smedlund, A., & Toivonen, M. (2007). The role of KIBS in the IC development of regional clusters. *Journal of Intellectual Capital, 8*(1), 159–170. https://doi.org/10.1108/14691930710715114

Spohrer, J., & Maglio, P. P. (2008). The emergence of service science: Toward systematic service innovations to accelerate co-creation of value. *Production and Operations Management, 17*(3), 238–246. https://doi.org/10.3401/poms.1080.0027

Starbuck, W. (1992). Learning by knowledge intensive firms. *Journal of Management Studies, 29*, 713–740. https://doi.org/10.1111/j.1467-6486.1992.tb00686.x

Strambach, S. (2001). Innovation processes and the role of knowledge-intensive business services (KIBS). In K. Koschatzky, M. Kulicke, & A. Zenker (Eds.), *Innovation networks: Concepts and challenges in the European perspective* (pp. 53–68). Physica-Verlag HD. https://doi.org/10.1007/978-3-642-57610-2_4

Strambach, S. (2010). Knowledge-intensive business services (KIBS) – On the way towards a knowledge processing and producing industry. In P. Cooke, C. De Laurentis, S. MacNeill, & C. Collinge (Eds.), *Platforms of innovation dynamics of new industrial knowledge flows* (pp. 170–204). Edward Elgar.

Sweeney, J. C., Soutar, G. N., & McColl-Kennedy, J. R. (2011). The marketing practices-performance relationship in professional service firms. *Journal of Service Management, 22*(3), 292–316. https://doi.org/10.1108/09564231111136845

Tether, B. S., & Hipp, C. (2002). Knowledge intensive, technical and other services: Patterns of competitiveness and innovation compared knowledge intensive, technical and other services. *Technology Analysis & Strategic Management, 14*(2), 163–182. https://doi.org/10.1080/09537320220133848

Tether, B. S., & Tajar, A. (2008). The organisational-cooperation mode of innovation and its prominence amongst European service firms. *Research Policy, 37*(4), 720–739. https://doi.org/10.1016/j.respol.2008.01.005

Toffler, A. (1980). *The third wave*. Morrow.

Toivonen, M. (2004). Expertise as business. Long-term development and future prospects of knowledge-intensive business services (KIBS) [Helsinki University of Technology]. In *Doctoral dissertation series* (Vol. 2). http://lib.tkk.fi/Diss/2004/isbn9512273152/

Toivonen, M., & Tuominen, T. (2009). Emergence of innovations in services. *Service Industries Journal, 29*(7), 887–902. https://doi.org/10.1080/02642060902749492

Tseng, C.-Y., Pai, D. C., & Hung, C.-H. (2011). Knowledge absorptive capacity and innovation performance in KIBS. *Journal of Knowledge Management, 15*(6), 971–983. https://doi.org/10.1108/13673271111179316

Tuominen, T., & Toivonen, M. (2011). Studying innovation and change activities in KIBS through the lens of innovative behaviour. *International Journal of Innovation Management, 15*(02), 393–422. https://doi.org/10.1142/S1363919611003209

Vargo, S. L., & Lusch, R. F. (2004). Evolving to a new dominant logic for marketing. *Journal of Marketing, 68*(1), 1–17. https://doi.org/10.1509/jmkg.68.1.1.24036

Vargo, S. L., & Lusch, R. F. (2008). Service-dominant logic: Continuing the evolution. *Journal of the Academy of Marketing Science, 36*(1), 1–10. https://doi.org/10.1007/s11747-007-0069-6

Vargo, S. L., Maglio, P. P., & Akaka, M. A. (2008). On value and value co-creation: A service systems and service logic perspective. *European Management Journal, 26*(3), 145–152. https://doi.org/10.1016/j.emj.2008.04.003

Werner, R. (2001). Knowledge-intensive business services in the Oulu Region. Business development and geographical linkage. In M. Toivonen (Ed.), *Growth and significance of knowledge intensive business services*. Uusimaa T&E Centre's Publications.

Windrum, P., & Tomlinson, M. (1999). Knowledge-intensive services and international competitiveness: A four country comparison. *Technology Analysis & Strategic Management, 11*(3), 391–408. https://doi.org/10.1080/095373299107429

Wood, P. (2002). Knowledge-intensive services and urban. *Urban Studies, 39*(5–6), 993–1002. https://doi.org/10.1080/00420980220128411

Wood, P. (2006). The regional significance of knowledge-intensive services in Europe. *Innovation: The European Journal of Social Science Research, 19*(1), 51–66. https://doi.org/10.1080/13511610600607718

Wood, P. A., Bryson, J., & Keeble, D. (1993). Regional patterns of small firm development in the business services – Evidence from the United-Kingdom. *Environment and Planning A, 25*, 677–700. https://doi.org/10.1068/a250677

Wyrwich, M. (2013). The role of regional conditions for newly emerging KIBS industries in the face of radical institutional change. *European Planning Studies, 21*(11), 1760–1778. https://doi.org/10.1080/09654313.2012.753694

Yam, R. C. M. M., Lo, W., Tang, E. P. Y. Y., & Lau, A. K. W. W. (2011). Analysis of sources of innovation, technological innovation capabilities, and performance: An empirical study of Hong Kong manufacturing industries. *Research Policy, 40*(3), 391–402. https://doi.org/10.1016/j.respol.2010.10.013

Zięba, M. (2015). Innowacyjność firm z sektora wiedzochłonnych usług biznesowych. *Studia i Prace Kolegium Zarządzania i Finansów/Szkoła Główna Handlowa, 145*, 163–175.

Zieba, M., Bolisani, E., & Scarso, E. (2016). Emergent approach to knowledge management by small companies: Multiple case-study research. *Journal of Knowledge Management, 20*(2), 292–307. https://doi.org/10.1108/JKM-07-2015-0271

KIBS Companies and Their Importance for Economy and Innovation

4

4.1 Review of the Literature on KIBS Companies

4.1.1 Phases and Areas of Research on Companies from the Knowledge-Intensive Business Services Sector

Research on companies from the KIBS sector has a relatively short history. The term "knowledge-intensive business services" was introduced to the literature only in the 1990s and from that moment they began to be analysed (Miles et al., 1995). This relatively short period of research on KIBS firms was divided by Muller and Doloreux (2009) into two phases:

1. *Phase one:* is a period of theoretical considerations regarding the selection of the KIBS sector as a separate sector in the economy.
2. *Phase two:* is a period of conducting various types of empirical research aimed at answering two questions: (a) do KIBS companies introduce innovations? (b) do KIBS companies introduce innovations in a different way than companies from the manufacturing sector?

According to these authors, the thematic scope of research on companies from the KIBS sector covers three main research trends: knowledge and KIBS companies, innovation and KIBS companies and spatial proximity and KIBS companies. Primary publications focus on emphasizing the importance of knowledge in KIBS services, as well as the interaction and exchange of knowledge between KIBS and its clients (Hauknes & Antonelli, 1999; Windrum & Tomlinson, 1999). The second line of research concerns the broadly understood issue of innovation, both in the context of supporting customer innovation (den Hertog, 2000)and creating innovations by companies from the KIBS sector (Larsen, 2000). The third stream of research is devoted to the issues of the location of KIBS companies—to what extent this location and spatial proximity affect the development of this type of

M. Zieba, *Understanding Knowledge-Intensive Business Services*, Knowledge Management and Organizational Learning 10,
https://doi.org/10.1007/978-3-030-75618-5_4

organization or the quality of services they offer (O'Farrell & Moffat, 1995; Wood, 2002).

Similar research areas concerning the KIBS sector were distinguished by Scarso (2015), who, on the basis of an analysis of articles about KIBS companies in the Scopus, Business Source Premier (BSP) and EconLit databases, proposed a division into the following four research areas:

- *KIBS companies and innovations:* research devoted to identifying the differences between innovations introduced by KIBS companies and those introduced by manufacturing companies, as well as showing that KIBS companies introduce not only technological innovations, but also innovations of a different type and use different innovation schemes;
- *KIBS companies and knowledge:* research on the cognitive aspects of KIBS companies, the impact of knowledge management on the activities of these companies, as well as knowledge transfer processes between KIBS and its customers, the ability to absorb and protect knowledge. Research in this area is still insufficient and fragmentary, especially considering the importance of knowledge in the activities of KIBS companies;
- *KIBS companies and space:*[1] research on the spatial location of KIBS companies, conducted mainly by geographers and economists combining the development of the KIBS sector with the development of regions or even entire economies;
- *KIBS companies and business aspects:* research on various business issues related to the functioning of KIBS companies, such as, for example, marketing or human resource management, as well as co-creating KIBS services in cooperation with clients. This line of research—similarly to the current of research on KIBS companies and knowledge—is still relatively underdeveloped and requires more attention from researchers.

In turn, the author of this book distinguishes chronologically three phases of research on companies from the KIBS sector (Zieba, 2013). The first phase concerned mainly the identification of KIBS companies and related theoretical aspects. The characteristic features of KIBS companies, indicated from the beginning of research on these companies, were their high level of innovation, as well as a significant contribution to the development of many economic sectors and the entire economy (Strambach, 2001; Muller & Zenker, 2001). The second phase of research was devoted to providing empirical evidence for the uniqueness of KIBS. Analyses were carried out on their innovative activities (Freel, 2006) or their interactions with customers (den Hertog, 2000; Toivonen, 2004), and the empirical foundations of the influence of KIBS companies on economic development were developed (Baláž, 2004; Thomi & Böhn, 2003; Wood, 2006). The third phase of research, which continues until today, includes an in-depth analysis of various factors and conditions related to the activities of these companies and their relations with the broadly

[1] In this context, space is understood as the environment (location) of KIBS companies.

understood environment (customers, partners, suppliers, R&D institutions, competitors, etc.).

In order to fulfil the methodological rigour of the study, the approach of systematic literature review was applied. Systematic literature review offers a systematic and transparent way for collecting, synthesizing and evaluating the findings of studies on a particular topic or question (Jesson et al., 2011, p. 104). The aim is to eliminate the bias resulting from the single studies and other types of reviews. The methodology for systematic literature review is derived from an evidence-based approach, initially used only in medicine, and then transferred to the field of organization and management science. It requires an unambiguous presentation of the research procedure, thanks to which it can be repeated by other researchers (Tranfield et al., 2003).

A systematic review of the literature was carried out in several stages:

1. defining the purpose of the review and formulating research question;
2. preparation of the protocol;
3. literature search and retrieval;
4. data organization and item selection;
5. synthesis and analysis of results;
6. interpretation of results and reporting conclusions (Petticrew & Roberts, 2006).

In the first stage, the aim of the literature review was determined, which was to perform a quantitative and qualitative analysis of publications on knowledge-intensive business services, and two research questions were formulated: (1) what research areas regarding knowledge-intensive business services are discussed in the literature? (2) what are the main achievements in each research area? Then, a research protocol was developed, which defined the theoretical background, the strategy of searching for literature together with terms and sources, selection of the research subject and selection criteria, as well as the strategy of data extraction and their analysis and synthesis (Orłowska et al., 2017).

For the analysis, publications from three full-text databases were selected: Business Source Premier (EBSCO), Scopus (Elsevier) and Web of Science (Thomson Reuters). It was decided to choose several databases at the same time, because each of them has a different reach and includes different publishers. These three specific databases were selected due to the fact that they cover the analysed years, contain magazines on broadly understood management and economics and also enable precise searching of articles in accordance with the assumed criteria. After selecting the databases, they searched for publications that contained the terms knowledge-intensive business services or KIBS—either in the title or abstract, or in keywords provided by the authors (if this option was available in the search engine). The number of publications found in individual databases is presented in Table 4.1.

As can be seen in the table above, the most publications were found in the Scopus database (2731), and the least in the Business Source Premier database (810).

In the next step, data organization and item selection took place. At this stage, the results were narrowed down to scientific articles published only in peer-reviewed

Table 4.1 Number of publications found in each database

Database	KIBS	Knowledge-intensive business services	Total
Web of Science	704	576	1280
Business Source Premier	376	434	810
Scopus	1312	1419	2731
Total	2392	2429	4821

Source: Own elaboration

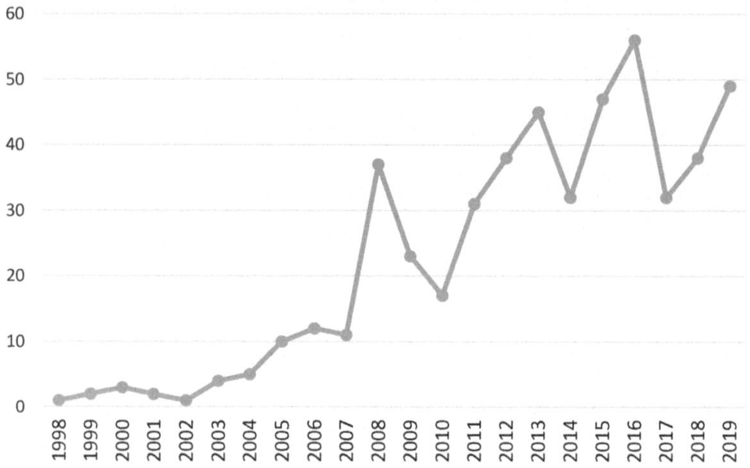

Fig. 4.1 Number of articles on KIBS published in 1998–2019 (Source: Author's own illustration)

journals, book chapters and post-conference publications, editorial introductions and other publications not belonging to the selected category were omitted. Academic peer-reviewed articles are considered the best source of data for systematic review (Jesson et al., 2011, s. 113). The list was also limited to publication only in English, which is considered the language of the congress. Articles published by the end of 2019 were analysed. Then all articles found in individual databases were merged, duplicates were searched, as well as articles not related to the analysed subject (e.g. in the field of medicine, history, etc.). After that, the so-called cleaning the database, a total of 496 studies were received.

The next stage included the synthesis and qualitative and quantitative analysis of individual articles, initially on the basis of titles and abstracts, and then—full versions of publications. The results are interpreted below.

The first of the analysed quantitative criteria was the number of articles devoted to KIBS published in particular years. As can be seen in Fig. 4.1, this number has been steadily increasing (despite some deviations) from 1998, when only 1 article on KIBS was published, to 2016, when 56 articles on this subject were published. After that year, there has been a decline to 32 articles in 2017. In the last analysed year, 2019, the number of published articles reached 49.

Another quantitative criterion was the number of articles published in individual journals. The "Service Industries Journal", which published as many as 36 articles about KIBS, turned out to be a definite leader in this respect. It was followed by "Regional Studies" (25 articles), "International Journal of Services, Technology and Management" (15 articles), "Industry and Innovation" (14 articles) and "Research Policy" (13 articles). Detailed information is summarized in Table 4.2, which includes all journals in which at least two articles were published in the analysed period.

In the next stage, after a detailed analysis of the publications, several research areas were created that these publications concerned, namely: KIBS sector companies and innovations, KIBS sector companies and their interactions with the environment, KIBS sector companies and knowledge and related processes, KIBS versus human resources and other areas (this category included diversified individual issues that were hardly covered in the analysed studies). This analysis shows that KIBS companies are examined from various perspectives (macroeconomic—at the sector level—and microeconomic—at the level of companies and phenomena occurring in them) and in relation to various areas of their operation (e.g. innovation, geographic distribution or interaction with other entities). Selected publications are listed in Table 4.3[2] concerning individual research areas, and then the main conclusions resulting from the presented research are described.

In the following sections, particular areas of the analysis of KIBS sector companies will be described, along with the main implications of research on these companies.

4.1.2 Companies from the Knowledge-Intensive Business Services Sector and Innovations

In the initial period of research, KIBS companies were perceived mainly through the prism of their impact on the innovation of their clients (Wood et al., 1993). Researchers focused on factors that increased the impact of KIBS companies on their clients' innovation capacity. Gradually, however, in the literature on KIBS companies, there was a change in their perception—from entities supporting or improving innovation (in the manufacturing sector) (Aslesen & Isaksen, 2007a; Toivonen, 2004) to co-creators and creators of innovation (Wong & He, 2005; Asikainen, 2013). Moreover, as the research shows, innovative activity in the services sector, in particular in the sector of knowledge-intensive business services, differs from innovative activity in production organizations (Tuominen & Toivonen, 2011). It has to do with less formal divisions into research and development and production activities and less popularity of innovations based on technologies implemented by KIBS companies. Other innovation discrepancies between KIBS and manufacturing companies are as follows:

[2]These publications were selected because they largely concerned companies from the KIBS sector.

Table 4.2 Number of articles on KIBS in selected journals (with at least 2 articles on KIBS published in the analysed period)

Journal	No. of articles
Service Industries Journal	36
Regional Studies	25
International Journal of Services, Technology and Management	15
Industry and Innovation	14
Research Policy	13
Service Business	12
Industrial Marketing Management	9
Knowledge Management Research and Practice	9
Amfiteatru Economic	8
European Planning Studies	8
International Journal of Technology Management	8
Journal of Knowledge Management	8
International Journal of Innovation Management	8
Economia Politica	6
Economics of Innovation and New Technology	6
Journal of Evolutionary Economics	5
Journal of Business Research	5
Engineering Economics	4
Innovation: Management, Policy & Practice	4
Journal of Economic Geography	4
Technological Forecasting and Social Change	4
Technovation	4
Journal of the Knowledge Economy	4
Sustainability	4
International Journal of Entrepreneurship and Innovation Management	3
Journal of International Entrepreneurship	3
Entrepreneurship & Regional Development	3
Geoforum	3
Journal of Intellectual Capital	3
R&D Management	3
Papers in Regional Science	3
Growth and Change	3
International Business Review	2
Kybernetes	2
Journal of Business and Industrial Marketing	2
Journal of Product Innovation Management	2
Foresight	2

Source: Own elaboration

- KIBS companies have a higher rate of innovation than manufacturing companies, but manufacturing companies "innovations are more related to R&D than KIBS companies" innovations.

Table 4.3 Review of research on KIBS sector companies

Research areas	Authors
KIBS companies and innovations	
Innovation of KIBS companies, R&D activities and protection of innovation	Czarnitzki and Spielkamp (2003), Asikainen (2013), Wong and He (2005), Amara et al. (2008), Miozzo and Grimshaw (2005), Corrocher et al. (2013), Miozzo et al., (2016), Tuominen and Toivonen (2011), Basilioa et al. (2019), Cabigiosu and Campagnolo (2019), Chichkanov et al. (2019), Desyllas et al. (2018), Doloreux et al. (2018), Doloreux et al. (2019a), Figueiredo and de Matos Ferreira (2019), Figueiredo et al. (2020), and Shearmur and Doloreux (2019)
Innovativeness of companies—clients	Aarikka-Stenroos and Jaakkola (2012), Yam et al. (2011), Aslesen and Isaksen (2007b), Toivonen (2004), and Wyszkowska-Kuna (2017a, 2017b)
KIBS company's approaches, strategies and types of innovation	den Hertog (2000), Freel (2006), Miles (2008), Amara et al. (2009), Corrocher et al. (2009), Rodriguez et al. (2015), Doloreux et al. (2019b), and Zieba et al. (2017)
KIBS companies and regional and national innovation systems	Muller and Zenker (2001), Koch and Stahlecker (2006), Bettiol et al. (2012), Corrocher and Cusmano (2014), and Chung (2002)
KIBS companies and their interactions with the environment	
Spatial proximity and regional concentration; territorial servicization; territorial knowledge dynamics	Doloreux et al. (2010), Koch and Stahlecker (2006), Antonietti et al. (2013), Deza and Lopez (2014), Figueroa-Armijos (2019), Lafuente et al. (2017, 2019), Qi et al. (2019), Werner and Strambach (2018), and Zenka et al. (2017)
Development of cities, economic sectors and regions	Aslesen et al. (2004), Ciriaci et al. (2015), Bujdosó et al. (2016), Wood (2006), Alvarez-Gonzalez and Gonzalez-Morales (2014), Borodako et al. (2015), Borodako et al. (2014), Desmarchelier et al. (2013), Alshubiri (2016), Brenner et al. (2018), Busu and Busu (2017), Chung and Tseng (2019), Figueiredo et al. (2019), and Yum (2019)
Clusters	Smedlund and Toivonen (2007) and Bocquet et al. (2016)
Networks and network strategies	Paiola et al. (2013), Siikonen et al. (2011), and Braga et al. (2017)
KIBS development policy	Viljamaa et al. (2010), Toivonen (2007), and Siikonen et al. (2011)
Internationalization of KIBS companies, export activities	Rodríguez and Nieto (2010), Bettiol et al. (2012), Peiker et al. (2012), Abecassis-Moedas et al. (2012), Braga et al. (2018), Doloreux et al. (2019a), Kamp and Ruiz de Apodaca (2017), Mol and Brandl (2018), Müller and Franz

(continued)

Table 4.3 (continued)

Research areas	Authors
	(2019), Rodríguez et al. (2018), and Wyszkowska-Kuna (2017a, 2017b)
Links with science and technology parks, universities	Calvo et al. (2017), Johnston and Huggins (2017), and Lee and Miozzo (2019)
Co-production and interactions with clients	Figueiredo et al. (2020), Hu et al. (2018), Kuula et al. (2018), Mustak (2019), and Zieba and Konczynski (2019)
KIBS sector companies and knowledge and related processes	
Knowledge absorption capacity	Tseng et al. (2011) and Koch and Strotmann (2008)
Knowledge externalities	Schmidt (2015), Rodriguez (2014), and Fernandes and Ferreira (2013)
Knowledge exchange (with clients)	Landry et al. (2012), Hu et al. (2018), Hu (2017), and Teirlinck (2018)
Expert knowledge	Aslesen and Isaksen (2007b)
Demand for knowledge	Antonelli (2017)
Knowledge management (KM); KM strategies; knowledge transfer	Bettiol et al. (2011), Palacios-Marques et al. (2011), Lara et al. (2012), Mangiarotti (2012), Zieba and Zieba (2014), Bolisani et al. (2016b), Bolisani et al. (2016c), Zieba et al. (2016), and Alexandru et al. (2019)
Protection of knowledge and intellectual capital	Bolisani et al. (2013), Päällysaho and Kuusisto (2008), and Päällysaho and Kuusisto (2011)
KIBS sector companies and human resources	
Required skills/level of education and competences; Human Resources Management	Consoli and Elche-Hortelano (2010), Toivonen et al. (2008), and Wojtczuk-Turek (2017)
Entrepreneurial behaviour	Ab. Majid and Awang (2016)
Other areas (examples)	
Sources of competitiveness of KIBS companies	Javalgi et al. (2011) and Xin et al. (2013)
Productivity, efficiency and growth of KIBS companies	Musolesi and Huiban (2010), Campisi et al. (2019), and Corsi et al. (2019)
Nature of KIBS services	Cabigiosu et al. (2015)
Trust	Scarso and Bolisani (2012) and Growe (2019)
Marketing	Bolisani et al. (2016a)
Sustainable development; sustainability	Bumberová and Milichovský (2019) and Lafuente et al. (2018)
Value and factors determining purchase of KIBS service	Heikka et al. (2018), Heikka and Mustak (2017), Heikka and Nätti (2018), and Heim et al. (2018)

- KIBS companies are characterized by a higher intensity of human capital, training expenditure, innovation expenditure and expenditure on research and development than manufacturing companies.

- KIBS and manufacturing companies have similar innovation goals.
- KIBS companies cooperate with foreign partners in the area of innovation less frequently than production companies (Wong & He, 2005).

Another aspect of the research on the innovativeness of KIBS companies was to determine what types of strategies and innovations these companies implement. Rodriguez et al. (2015) analysed the factors determining innovation strategies in KIBS enterprises, examining the relationship between these strategies and two aspects of innovation: the type of innovator (internal, collaborative or external) and the degree of innovation openness (dependence on internal or external sources of information). The results of the empirical analysis showed a clear relationship between the two identified innovation strategies and both the degree of innovation openness and the type of innovator. Conversely, Amara et al. (2009) identified potential factors that could affect the types of implemented innovations and examined their impact. Other researchers distinguished four KIBS company profiles: an interactive innovation profile, a production innovation profile, a conservative innovation profile and a techno-organizational innovation profile (Corrocher et al., 2009). Among the factors influencing the selection of a particular innovation profile, the company's strategy turned out to be the most important, followed by its size, customer location and employee training. The innovativeness of KIBS firms has also been analysed with regard to the knowledge it can potentially be supported with and approaches towards obtaining it (Zieba et al., 2017). According to the authors, companies may manifest a passive or an active behaviour in its obtaining and they can also rely more on internal (employees, in-house R&D, internal documents, etc.) or external (clients, suppliers, service providers, universities, etc.) sources of knowledge (Zieba et al., 2017). External knowledge search strategies were also examined in a study by Doloreux et al. (2019a). In some recent studies, also the issues of R&D in KIBS firms have been discussed, e.g. Basilioa et al., 2019; Doloreux et al., 2018. Additionally, the aspect of innovation has been explained in a more complex contest, e.g. what are its drivers (Chichkanov et al., 2019), how it is related to growth (Cabigiosu & Campagnolo, 2019), value capturing (Desyllas et al., 2018) or co-creation (Figueiredo et al., 2020).

Another important stream of research on KIBS companies is their place in regional and national innovation systems. The innovation system—both regional and national—can be defined as a set of various actors (entities), interconnected and influencing the innovation processes in a given area/country (Chung, 2002). In other words, it is a system of entities, such as various types of organizations (including KIBS companies) and institutions (e.g. research and development institutes, universities or public institutions), which as a result of synergy arise in a specific territory and contribute to increasing the ability to create, absorb and diffuse innovation in a given area (regional/national). Research on the relationship of KIBS companies with innovation systems shows that KIBS companies are a decisive element of innovation-driven regions, and their scarcity and slow growth are features that clearly characterize poorly efficient innovation systems (Corrocher & Cusmano, 2014). Moreover, beyond the macroeconomic implications, newly founded KIBS firms are believed to play an important role in regional production and innovation systems. As the founders of start-ups in the early stages of their development make the most use of regional resources, and KIBS companies gain knowledge through an interactive process that takes place at the time of service provision, the interdependence between regional economic, technological and institutional organizations and newly founded KIBS companies (Stahlecker & Koch, 2004; Koch & Stahlecker, 2006). In general, researchers agree that both KIBS companies have a positive impact on the development of innovation systems, and the development of KIBS companies, especially new ones, depends on regional innovation systems.

Research on the innovativeness of KIBS companies has gradually evolved from the analysis of issues related to the innovativeness of companies—clients to researching the innovativeness of the KIBS companies themselves and their relations with innovation systems (regional/national). In fact, researchers agree that KIBS companies introduce innovations in a different way than, for example, manufacturing companies, and that apart from their own innovation (often at a high level, although it is not a universal principle), they contribute to the innovation of their customers and the wider environment. It should also be noted that with time more complex studies were conducted, examining not only the innovativeness of KIBS firms and their clients, but also its links with other concepts and influence on other research areas.

4.1.3 Companies from the Knowledge-Intensive Business Services Sector and Their Interactions with the Environment

Another area of analysis of KIBS companies is their broadly understood interactions with the environment. Companies from the KIBS sector tend to concentrate regionally and influence the development of the areas in which they are located (see: Aslesen et al., 2004; Bujdosó et al., 2016; Ciriaci et al., 2015; Wood, 2006). Moreover, the increased availability of KIBS services has a positive effect on productivity and other services, such as tourism (Alvarez-Gonzalez & Gonzalez-Morales, 2014). Interestingly, KIBS services are intensively used, among others by companies from the tourism sector (Borodako et al., 2015, 2014). Overall, KIBS companies can be seen as engines of economic development and function as a substitute for capital accumulation (Desmarchelier et al., 2013). In some recent studies, a phenomenon called territorial servicization has been examined (Figueroa-Armijos, 2019; Lafuente et al., 2017, 2019). This phenomenon can be explained in the following—"the development of knowledge-intensive business service (KIBS) sectors in certain territories contributes to rebuild the competitive advantage of manufacturing businesses" (Lafuente et al., 2019, p. 313).

KIBS companies are also an interesting and important field of research also for researchers dealing with clusters and networks. Networking is one of the hallmarks of KIBS companies that usually interact with their customers, suppliers or even competitors in complex interactions and choose different networking strategies.

Due to the significant positive influence of KIBS companies on other entities and economic areas, there is a trend of research on the KIBS sector promoting solutions supporting its development. Some researchers even provide guidelines for formulating a policy to support KIBS companies (Viljamaa et al., 2010; Toivonen, 2007; Siikonen et al., 2011). It should be noted, however, that the existing policies to support KIBS companies concerned only the aspect of innovation and the perception of the importance and the need to support KIBS companies in this area. Moreover, all the publications mentioned above were about Finland and supporting the innovation of KIBS companies in that country. Activities in the area of support programs are few and they focus mainly on the aspect of supporting the innovation of KIBS companies. The reasons for this state of affairs may be the commonly existing belief in the importance of innovation for the development of the economy, as well as the relative ease in measuring the results of innovative KIBS companies. If there was a policy of supporting KIBS companies in the area of knowledge exchange or network activities, its effects would be more difficult to measure than in the case of innovation.

KIBS companies have also been recently researched in the context of their internationalization. There are two most popular ways of internationalizing KIBS companies: (1) through foreign direct investment in some form (e.g. acquisitions, joint venture with foreign partners) and (2) through KIBS staff travelling to other countries, mainly due to the need for close cooperation between KIBS and the client foreign. In addition, KIBS companies can internationalize gradually or from scratch (*born global*) (Bettiol et al., 2012). So far, the internationalization of KIBS companies has been analysed in the context of factors influencing it, such as the size, location and innovativeness of the company, as well as their potential contribution to internationalization of countries (e.g. Kamp & Ruiz de Apodaca, 2017). Research results indicate that KIBS companies that establish cooperation have easier access to international markets and improve their innovative capacity (Rodríguez & Nieto, 2012). Consequently, the cooperation has a direct and indirect relationship with the internationalization of KIBS companies. The specific characteristics of services, especially in terms of knowledge and relationships with customers, significantly affect the spatial results of international expansion (Peiker et al., 2012). The country of origin with a particular institutional environment also influences the possibilities and goals of expansion. There are also studies discussing KIBS export strategies (Doloreux et al., 2019b). Summing up, it can be said that the internationalization of KIBS companies has so far been analysed by a still small number of researchers and it is undoubtedly an interesting research area for further exploration.

A relatively new areas of KIBS firms analysis are: links with science and technology parks, universities and co-production and interactions with clients. More and more studies concern how KIBS firms can be supported with knowledge from other entities, e.g. universities. Also, there are a few studies on the joint production of KIBS firms with their customers and the characteristics of their interactions. This seems to be another promising field of study.

4.1.4 Companies from the Knowledge-Intensive Business Services Sector and Their Basic Resource

Knowledge as a key resource of KIBS companies was analysed in research on the KIBS sector from various perspectives and in various contexts. Some studies looked at how knowledge is absorbed by KIBS companies (Tseng et al., 2011; Koch & Strotmann, 2008) or the expertise of these companies and their role in creating innovation (Aslesen & Isaksen, 2007b), others, in turn, focused on research and development and the exchange of knowledge with customers (Landry et al., 2012; Hu, 2017; Hu et al., 2018; Teirlinck, 2018). A relatively new area is research on the ways in which KIBS companies manage their knowledge and the types of practices they implement for this purpose. For example, Bettiol et al. demonstrated that all knowledge management strategies (knowledge codification, personalization and knowledge creation) implemented by KIBS are positively correlated with market expansion by these companies (Bettiol et al., 2011). The study was conducted on a sample of over 150 Italian KIBS specializing in design and communication. Knowledge management strategies selected by KIBS companies were also analysed in the context of their characteristics and division into an emergent strategy and a deliberate strategy (Bolisani et al., 2016b, 2016c; Zieba et al., 2016).

Another research area of KIBS companies is the relationship between knowledge management (KM) and human capital. Palacios-Marques et al. examined how knowledge management practices affect the development of human capital in KIBS companies. Researchers conceptualized KM practices in six dimensions[3] and tested the relationship between each dimension and human capital. The study was conducted in Spanish companies belonging to the biotechnology and telecommunications industries and shows that the introduction of KM practices has a positive impact on the development of human capital. Firms that adopt continuous learning practices and develop an innovative culture that encourages R&D, according to researchers, improve the skills and abilities of their human capital (Palacios-Marques et al., 2011).

Attempts were also made to determine how knowledge management influences the company's results. In a study conducted on a sample of 122 KIBS companies employing over 100 employees, the relationship between GM and the results (tangible and intangible) achieved by these companies was checked. As shown, material results are positively correlated with knowledge resources and negatively correlated with strategic planning processes. Additionally, tangible performance also depends on management systems. Interestingly, knowledge management turns out to have a greater impact on the material results of the surveyed companies than on non-material ones, which is a surprising conclusion, rarely found in the literature. In

[3]These dimensions include: (1) orientation on the development, transfer and protection of knowledge as well as management of knowledge resources and knowledge flows; (2) continuous learning in the organization; (3) understanding the organization as a global system; (4) developing an innovative culture that encourages research and development projects; (5) people-based approach; (6) competence development and competence-based management (Palacios-Marques et al., 2011).

conclusion, if a company wants to improve its performance, it must be able to constantly learn and apply knowledge, anticipating changes in the market (Lara et al., 2012). A relationship was also demonstrated between the introduced KM practices and the propensity to innovate (Mangiarotti, 2012) and the impact of two key success factors in knowledge management (i.e. leadership and management support and motivational practices) on the innovativeness of KIBS SMEs (Zieba & Zieba, 2014).

The last area of research related to knowledge and also relatively little researched concerns the protection of knowledge and intellectual property in KIBS companies. KIBS companies do not seem to use methods of protecting their knowledge (especially formal ones) and in most cases they do not have a planned strategy of protecting their own knowledge, which may hinder this protection (Bolisani et al., 2013). Partially different results came from studies by Finnish researchers, in which KIBS companies do not always use formal means to protect their intellectual property, but often use informal protection mechanisms (Päällysaho & Kuusisto, 2008, 2011). This is especially true for small organizations for which these informal methods are simple, controllable and economical to use and are often an integral part of their daily organizational routine.

4.1.5 Companies from the Knowledge-Intensive Business Services Sector: Other Research Areas

Companies from the KIBS sector are becoming an area of interest also in terms of human resources that are employed in such organizations. The analysis of the skills of people employed in KIBS companies shows that these skills are diverse (Consoli & Elche-Hortelano, 2010). This differentiation is most evident when comparing KIBS companies whose main activity is to maintain the infrastructure for reporting information and those that create new information. In companies of the first type, repetitive procedures dominate, and in companies of the second type—the formulation of problem-solving strategies around the current task. Other aspects were dealt with in their research by Toivonen et al., who investigated the development of specialist knowledge in KIBS companies and the ways in which these companies try to meet the challenges arising from this development (Toivonen et al., 2008). According to the researchers, firstly, knowledge of customers' business and mastery of the process of cooperation with customers play an increasing role in the provision of KIBS services, and secondly, knowledge becomes more multidisciplinary, i.e. less related to specific academic disciplines. In other words, in today's KIBS companies, expertise is not limited to single, well-defined occupations, but is increasingly multidisciplinary and reflects the blurring of professional boundaries. The ability to link your service to the customer value chain in a mutually beneficial way has become an integral part of the expertise of KIBS companies. The factors contributing to the development of this type of knowledge in KIBS companies include globalization, the universality of network activities and the development of information and communication technologies, generating a demand for expert services.

As can be seen from the example of the research areas in the field of human resource management discussed above, KIBS companies are an interesting field for research also in this aspect. This is due to the high dependence of this type of organization on the knowledge possessed and acquired by their employees. It should be assumed that a more detailed study of the characteristics of this knowledge and effective human resource management in KIBS companies may help such companies achieve better financial and non-financial results.

KIBS companies are also starting to be analysed as a specific type of company that requires a different management method. So there are publications on their productivity, efficiency and growth (Musolesi & Huiban, 2010; Campisi et al., 2019; Corsi et al., 2019), character of the offered services (Cabigiosu et al., 2015) or trust (Scarso & Bolisani, 2012; Growe, 2019). There are also a few recent publications on the areas such as sustainable development and sustainability (Bumberová & Milichovský, 2019; Lafuente et al., 2018) or value and factors determining purchase of KIBS service (Heikka et al., 2018; Heikka & Mustak, 2017; Heikka & Nätti, 2018; Heim et al., 2018).

- *On the basis of the literature review presented above, the following conclusions can be drawn. The most studied area of operation of KIBS companies is the innovative activity of these enterprises and their clients' companies. It was in this context that researchers first became interested in the KIBS sector and the companies operating in it. Later, the KIBS sector began to be analysed in relation to its interactions with the environment, and in particular its impact on the development of cities, regions or economic sectors—in a word, from a macro-economic perspective. Only recently have KIBS companies started to be analysed in terms of processes, activities and specificity of their functioning,* e.g. *human resources management or applied strategies. One of such areas of research is knowledge and related aspects, such as knowledge management, which are still relatively poorly understood in comparison with, for example, the issue of innovation or the interaction of KIBS companies with the environment. Although KIBS companies rely heavily on their knowledge, or on knowledge that they can reach and use, little is known about how they manage their knowledge and that of their clients. A completely unexplored area is the flow of knowledge between KIBS companies and their clients, which in a way determines the creation of the KIBS service.*

It should be borne in mind that the division into research areas (Table 4.3) is arbitrary, because many of the publications listed cover more than one area. For example, the study performed by Rodríguez and Nieto (2010) concerned the tendency of KIBS companies to internationalize and the factors that potentially influence this internationalization (innovative capacity and cooperation with other entities), and the study conducted by Smedlund and Toivonen (2007) examined the relationship between presence in the cluster and intellectual capital. Another example is a study by Pace and Miles (2019), who presented a study on KIBS interactions with clients on absorptive capacity building for environmental innovation (three various research areas). There are many such examples of dealing with several different aspects of KIBS operations. This shows the interdisciplinary nature of research on KIBS companies, which eludes the standard characteristics of service companies. The number of studies linking a variety of research areas related to KIBS is growing in time.

4.2 The Importance of the KIBS Sector for the Economy

A significant number of publications on the KIBS sector presented in the previous parts of this paper concern the growth of this sector and its importance for the development of the entire economy. Regions with strong KIBS sectors are characterized by the highest level of prosperity in Europe, and the presence of a strong KIBS sector positively affects the level of regional innovation (number of patents) (The European Cluster Observatory, 2009). KIBS companies have a positive impact on the competitiveness of the sectors of the economy that use them in their production processes, mainly increasing the productivity of these entities (Wyszkowska-Kuna, 2016).

The importance of the knowledge-intensive business services sector has both a quantitative and a qualitative dimension. When it comes to quantitative indicators confirming the growing importance of the KIBS sector, such indicators as employment in this sector or the added value it generates can be mentioned. Companies in this sector often employ well-educated specialists from various areas, and the number of people employed in enterprises from the KIBS sector in the European Union is systematically growing (Fig. 4.2).

Also in most individual European countries, over the years 2008–2016, the number of people employed in the KIBS sector increased (Table 4.4).

In almost all analysed countries, the number of people employed in the KIBS sector increased—the only exception is Greece, where the number of people employed in KIBS companies in 2016 was over 10% lower than in the base year 2008. This phenomenon can be explained by the crisis of the Greek economy. Due to financial problems, many companies could resign from using the services of KIBS companies, which in turn had to reduce the number of employees or even terminate their activities. This could have resulted in an outflow of highly educated workforce to other sectors of the economy and even beyond the country's borders. In Spain and Italy, the increase in employment in the KIBS sector was the lowest (by less than 5%) and in Turkey the highest (by 142.3%!). Interestingly, among all European Union countries for which all data are available, the highest employment growth in the KIBS sector was recorded in Poland—by almost 50%. Countries such as Switzerland (40.1%), the Czech Republic (37.5%) and Great Britain (31.9%) can also boast of high growth. The large increase in employment in the KIBS sector in Poland additionally justifies the importance of undertaking research on companies offering knowledge-intensive business services.

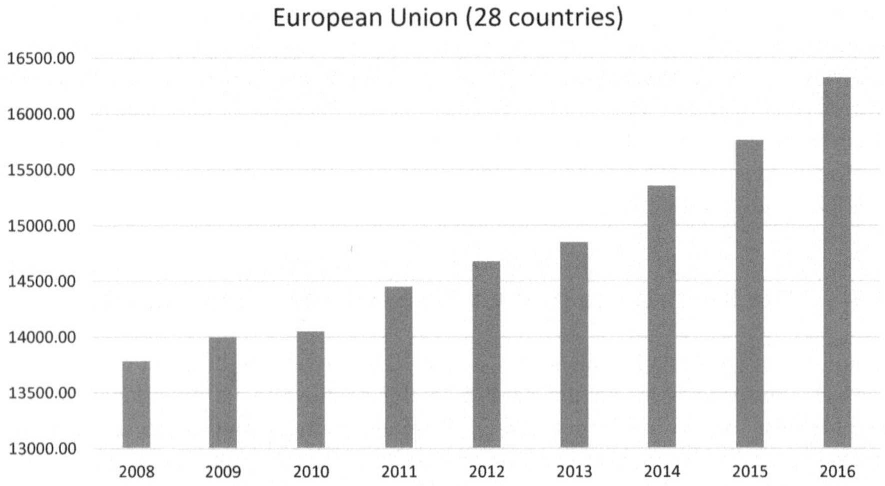

Fig. 4.2 The number of employees in the KIBS sector in the European Union (in thousands) according to the NACE Rev.2.0 classification (Source: Author's own illustration, based on data from Eurostat)

The importance of the KIBS sector also has its qualitative dimension. These companies are becoming more and more significant sources and channels of new knowledge (Miles, 2005, p. 40). As KIBS companies influence the activities of their clients, this translates into the operation of the entire economy. KIBS companies play many roles in the economy, which are characterized below:

- direct transfer of expert knowledge, i.e. the traditional model of consulting practice;
- sharing experiences, transferring experiences and ideas from one context to another;
- benchmarking, within which you can jointly develop a process of identifying and focusing on "good practices";
- knowledge brokering, combining various sources of knowledge and its users;
- diagnosing and solving problems, as well as helping users articulate and define specific needs in such a way that external resources and opportunities can be effectively used to meet them;
- acting as an agency for introducing changes, thanks to which it is possible to develop a company—a client using a neutral external perspective (Miles, 1999).

The growing importance of the KIBS sector and the companies operating in it is due to the following factors:

Table 4.4 The number of employees in the KIBS sector in individual European countries, according to the NACE Rev.2.0 classification

	2008	2009	2010	2011	2012	2013	2014	2015	2016	% increase
Austria	252.8	263.7	275.7	268	281.1	289.7	301	302.1	311.3	23.10%
Belgium	296.8	307.6	299.7	317.3	344.8	363.3	368.6	380.1	383.4	29.20%
Bulgaria	91.70[a]	94.20[a]	103.50[a]	96.40[a]	97.80[a]	104.70[a]	111.60[a]	135.50[a]	145.20[a]	–
Croatia	58.20[a]	57.80[a]	67.80[a]	75.80[a]	73.4	77.3	73.2	77.6	85.7	–
Cyprus	21.20[a]	20.90[a]	22.20[a]	24.30[a]	26.60[a]	25.70[a]	26.40[a]	27.10[a]	26.30[a]	–
The Czech Republic	235.5	247	254.1	260.4	267.4	288.6	289.5	301.4	323.8	37.50%
Denmark	202.2	207	210.1	204.1	205.4	206.7	210.2	223.7	234.8	16.10%
Estonia	25.20[a]	26.60[a]	27.30[a]	29.90[a]	30.20[a]	33.10[a]	37.40[a]	39.8	41.5	–
Finland	207.6	199.7	200.5	208.8	216.8	222.1	228.2	233	227.1	9.40%
France	1660.2	1697.1	1716.5	1874.4	1857.4	1820.3	1879.4	1914.6	1949.3	17.40%
Greece	252.6	254.2	236.7	232.6	235.3	218.8	218.1	225.9	222.60[a]	−10.6%[b]
Spain	1147.8	1121	1123	1115.1	1112.4	1085.2	1106.2	1142.7	1199.5	4.50%
The Netherlands	836.6	767.6	707.9	691.9	722.3	851.1	867.4	887.9	907.8	8.50%
Ireland	154.9	145.4	143.6	145.3	152.8	160.2	165.7	171.3	176.5	13.90%
Island	11.70[a]	11.90[a]	12.50[a]	12.70[a]	11.60[a]	11.80[a]	14.00[a]	15.20[a]	15.70[a]	–
Lithuania	34.00[a]	40.10[a]	43.80[a]	50.90[a]	50.50[a]	55.80[a]	59.30[a]	66.30[a]	68.10[a]	–
Luxemburg	15.70[a]	16.90[a]	17.80[a]	15.80[a]	20.2	21.40[a]	23.30[a]	22.6	22	–
Latvia	35.40[a]	27.20[a]	29.00[a]	31.20[a]	33.80[a]	42.60[a]	47.50[a]	47.40[a]	44.20[a]	–
Macedonia	–	–	–	15.5	18.5	14.50[a]	17.20[a]	17.20[a]	19.20[a]	–
Malta	6.90[a]	7.80[a]	8.10[a]	9.60[a]	10.40[a]	10.30[a]	12.10[a]	13.80[a]	14.30[a]	–
Germany	2555.7	2619.3	2674	2843.4	2818.2	2648.5	2679.6	2707	2845.7	11.30%
Norway	179.6	186.2	180.6	185.7	198	203	210.7	211.5	214.7	19.50%
Poland	504.9	566.2	571.1	591.9	648	662.2	708.8	726.6	756.9	49.90%
Portugal	186.30[a]	184.80[a]	179.10[a]	189.60[a]	172.00[a]	186.20[a]	223.90[a]	227.00[a]	238.20[a]	–
Romania	150.70[a]	161.00[a]	167.70[a]	184.60[a]	193.70[a]	202.20[a]	229.70[a]	248.80[a]	256.10[a]	–
Slovakia	93.30[a]	102.20[a]	102.3	101.5	100.8	100.5	112.4	118.7	125.7	–

(continued)

Table 4.4 (continued)

	2008	2009	2010	2011	2012	2013	2014	2015	2016	% increase
Slovenia	50.8	52.9	57.6	56	57.3	54.2	59.8	59.9	58	14.20%
Switzerland	361.4	380.6	377.9	387.1	415	437.3	466.7	483.8	506.5	40.10%
Sweden	482	478.4	491.5	519.3	532.6	542.4	564	584.7	609.7	26.50%
Turkey	–	438.3	465.5	481.3	580.5	634.3	794.3	921.1	1062	142.3%[c]
Hungary	188.3	175.4	173.5	178.2	185.2	203.7	210.2	222.7	227.6	20.90%
The Great Britain	2288.5	2462.8	2449.9	2452.4	2585.6	2707.3	2814.2	2910	3017.9	31.90%
Italy	1698.1	1653.5	1657.5	1650.6	1608.9	1638.4	1699.9	1715.3	1771.9	4.30%

Source: own study based on data from Eurostat

[a]Data not available for all subsectors

[b]For Greece, the end year is 2015

[c]For Turkey, the base year is 2009

- outsourcing[4] services from customers, although a related trend in offshoring services[5] could theoretically reduce the number of KIBS companies located in the European Union;
- technical/technological progress and growing demand for various types of technological knowledge, especially related to new technologies, such as information technology (IT);
- growing demand for specialist knowledge on social, administrative and regulatory issues—from understanding consumer and industrial markets to complying with and anticipating regulatory changes;
- factors related to internationalization and globalization of business activity;
- increasing emphasis on services and intangible elements of production and intangible products in the knowledge-based economy;
- developing markets for knowledge workers (PREST, 2005).

The first factor influencing the development of the KIBS sector is the outsourcing of services. Firms are increasingly outsourcing services that were originally internal operations. This is due to various reasons, the most significant of which are the philosophy of focusing on the core competencies of the organization and reducing internal costs in order to increase competitiveness (Huggins, 2011). Companies that offer specialized business services through outsourcing can expect economies of scale from having multiple customers, as well as efficiency and effectiveness from experiences and learning from different customers, the need to compete with other vendors, and other ways to improve their results. At the same time, the company (the customer) benefits from greater flexibility—contracts and changes between service providers can be easier than collaborating with the workforce employed in the organization. It should be remembered that there are transaction costs associated with seeking and using external service providers. Company–customer must have adequate absorption capacity, otherwise service benefits may be lost. Smaller companies may find it difficult to find funds to use KIBS services and be forced to use only their own internal resources. Not all activities may be outsourced, e.g. R&D often remains within the organization (Miles, 2005). Summing up, outsourcing is a factor conducive to the development of KIBS services, however, it has limitations such as the type of activity that may be outsourced or the costs associated with the purchase of services.

Another phenomenon influencing the development of the KIBS sector and related to outsourcing is offshoring. "Offshoring is an important form of acquiring resources for the service sphere, and includes services of varying complexity and added value,

[4]Outsourcing is a management strategy consisting in outsourcing to a specialized external company tasks that are not directly related to the core business profile of the company ordering, https://encyklopedia.pwn.pl/encyklopedia/outsourcing.html (date of access: 07.09.2018).

[5]Offshoring is the practice of basing a business or part of a business in a different country, usually because this involves paying less tax or other costs, https://dictionary.cambridge.org/pl/dictionary/english/offshoring (date of access: 12.10.2020).

such as call-centres on the one hand, and R&D or software development on the other" (Skórska, 2012, p. 142).

Progressive globalization, elimination of formal barriers to the flow of resources, as well as the possibility of reducing costs and achieving economies of scale make companies more and more willing to partially relocate their activities outside the country. KIBS companies benefit from this, as they can provide their services to foreign companies and can also transfer their activities to other countries. As a result, they can reduce costs and concentrate more on their core business and consequently achieve increased productivity and added value. Offshoring also has its drawbacks, to which it should be included:

- economic risk, related to, inter alia, with the exchange rate, international wage rates, etc.;
- influence of political factors (e.g. political stabilization of the country);
- strategic issues (e.g. adjusting the service to the client's needs, access to qualified resources);
- inability to engage in close interaction with the client;
- difficulties in building trust;
- technological limitations in the digitization of service activities;
- the risk of losing control over the company's key activities;
- risk of losing confidential data, information and knowledge (Skórska, 2012, pp. 145–146).

Technical progress is another important factor in the development of the KIBS sector. Since the 1990s, a rapid acceleration of the pace of technological progress can be observed, also referred to as the IT revolution or technological and information revolution, resulting from the emergence and development of information and communication technologies. The IT revolution changed the functioning of many types of service companies, especially those offering knowledge-intensive business services, because services are often intangible and largely saturated with knowledge and information—this makes them strongly related to ICT[6] (Wyszkowska-Kuna, 2016, pp. 45–46). Technological development not only changes the way of creating and delivering many KIBS services, but also helps to increase the demand for such services. Rapid technological progress and the proliferation of applications pose a huge problem for organizations that want to effectively use new technologies. It is necessary to acquire large amounts of new knowledge to keep up with these changes. KIBS companies, especially those in the T-KIBS category, can help organizations fill the gaps in their knowledge of new technologies and how to use them.

Another factor contributing to the growing demand for KIBS services is the growing demand for specialist social, administrative or regulatory knowledge.

[6]It should be remembered that this relationship applies to a different degree to different types of services and will be greater, e.g. in the case of computer services, and smaller in the case of, e.g., hairdressing services.

Specialist legal and accounting services have existed for a long time, but are now gaining in importance. For example, compliance with environmental regulations, health and safety standards and many similar issues is a major challenge for companies. The challenges are greater with companies operating in different countries and with different traditions. KIBS companies provide basic information, advice and brokerage services. They also help train the company's staff—the customer (especially the more technology-oriented KIBS companies). Other KIBS companies help clients to better understand the markets and cultures, consumers and other stakeholders they work with. Examples of such services are marketing research and public relations. Product proliferation and diversification of consumer demand resulted in a significant increase in demand for this type of services. Also in this case, companies operating in the international environment have a greater demand for knowledge about their environments and ways of functioning in them. Environmental regulation and social concerns related to environmental issues are important factors influencing the growth of environmental services, partly provided by KIBS companies (e.g. consultancy services on "clean" technologies and environmental law). Environmental issues can affect the operation of both KIBS companies, which are usually less energy-intensive than many other service companies (in particular in the area of transport, but also numerous physical services and services requiring large office infrastructures), and companies—customers who may be more willing to use the services of KIBS companies in the field of environmental consulting or "clean" technologies (Miles, 2005, p. 45).

Another factor influencing the development of the KIBS sector is related to the internationalization and globalization of business activities. As mentioned in the previous section, little is known about the internationalization of KIBS companies themselves, but one thing is certain—the advancing globalization affects the development and internationalization of business services (including KIBS), which in turn are becoming a source of further market integration and increased competition. In other words, on the one hand, globalization and internationalization increase the demand for KIBS services, and on the other hand, they make KIBS companies act internationally, following their customers and trying to better match their needs. The impact of globalization on the development of KIBS services includes on the increase in export opportunities, resulting from the elimination of trade development barriers, creating demand for new types of KIBS services (e.g. related to operating in an international environment), international trade in business products and services related to ICT technologies or harmonization of market regulations within the European Union (Skórska, 2012, p. 130). KIBS companies take up the challenge of internationalization for a variety of reasons:

- willingness to follow their customers in new operating environments;
- willingness to find new markets;
- the need to compete with foreign KIBS companies, which are entering the markets of an increasing number of countries (increasing such competitive pressure is fostered by trade liberalization agreements);

- willingness to acquire new knowledge, practices from other countries, and thus learn new things and improve (Miles, 2005, p. 46).

Globalization and internationalization for KIBS companies also come with constraints. One of the main limitations is the need for close (often geographically close) cooperation with the company—the client. Other limitations include: the national nature of the rights and competences of professionals working in KIBS companies, language and cultural differences, regulations at the national level, trade and economic barriers. However, despite potential barriers and limitations, KIBS services have an increasing share in access to capital, technology and new markets.

Another phenomenon favouring the development of the KIBS sector is the growing role of services and intangible elements of production and products. One of the important components of the development of the knowledge economy is the general growth of the "service" as the operating principle in economic life (sometimes referred to as the "servicization" of the economy). Businesses across all sectors recognize that their competitive advantage lies in the services they provide that their customers receive in the transaction, not in the specific goods they sell. This means paying more attention to the components of customer relationship services—from customer involvement in the selection and design of products, to sales and disposal. KIBS companies can support companies—clients in creating and delivering this type of services, therefore the demand for them will grow with the increasing degree of servicization of the economy.

The development of markets for knowledge workers is another factor contributing to the development of the KIBS sector. Knowledge workers constitute a specific group and can be defined as employees who can independently find, understand and apply knowledge in a given organization, and apply the acquired knowledge in their daily work, solving problems of their clients (Gottschalk, 2005, p. 27). KIBS companies are often a dream place of work for knowledge workers, because they offer them not only a satisfactory level of remuneration, but also the possibility of self-fulfilment in their work and a variety of professional experiences. Companies using the services of KIBS companies very often could not afford to employ such specialists, and thanks to using the services of KIBS, they can benefit from their knowledge.

Added to the above list are the growing importance of innovation and the challenges of increasing competition. The evolution towards a knowledge-based economy not only poses a new challenge for the competitiveness of companies, but also involves a change in both the nature of the organization and the way in which they create and implement their strategies (Huggins, 2011). Increasing dependence on the creation of wealth from intangible assets is making the global economy more fluid and volatile, and the ability to effectively access knowledge and combine new and existing knowledge has become crucial to the competitiveness of companies, regions and nations. In this context, KIBS companies support other organizations in their competitiveness on the market by reaching out to knowledge and combining new and existing knowledge.

4.3 Summary

In this chapter, a systematic literature review on KIBS companies has been presented. It can be clearly seen that this topic of research is quite diversified and analysed from a variety of perspectives. Since the creation of this term in 1995, many researchers have been exploring KIBS services and KIBS companies not only from the macro perspective of their influence on the economy, but also from the micro perspective of their operations and interactions between them and their customers and the broader environment. Taking into account the importance of the KIBS sector for this both regional and global environment, it is definitely important to help these companies in efficiency improvement, innovation management and other areas of operations. That is why, past and future research areas have been identified in this chapter, shading some lights on the research possibilities.

References

Aarikka-Stenroos, L., & Jaakkola, E. (2012). Value co-creation in knowledge intensive business services: A dyadic perspective on the joint problem solving process. *Industrial Marketing Management, 41*(1), 15–26. https://doi.org/10.1016/j.indmarman.2011.11.008

Ab Majid, M. N., & Awang, M. G. (2016). The implication of entrepreneurial behavior and network collaboration on firm innovativeness in knowledge intensive business services. *International Business Management, 10*(15), 2756–2767.

Abecassis-Moedas, C., Ben Mahmoud-Jouini, S., Dell'Era, C., Manceau, D., & Verganti, R. (2012). Key resources and internationalization modes of creative knowledge-intensive business services: The case of design consultancies. *Creativity and Innovation Management, 21*(3), 315–331. https://doi.org/10.1111/j.1467-8691.2012.00646.x

Alexandru, V. A., Bolisani, E., Andrei, A. G., Cegarra-Navarro, J. G., Martínez Martínez, A., Paiola, M., Scarso, E., Vătămănescu, E.-M., & Zieba, M. (2019). Knowledge management approaches of small and medium-sized firms: A cluster analysis. *Kybernetes, 49*(1), 73–87. https://doi.org/10.1108/K-03-2019-0211

Alshubiri, F. N. (2016). Analysis of economic development based on kibs interaction: Evidence from sultanate of Oman. *Actual Problems of Economics, 2016*(6), 73–83.

Alvarez-Gonzalez, J. A., & Gonzalez-Morales, O. M. (2014). The role of knowledge-intensive business services in Spanish local tourist production systems. *Tourism Economics, 20*(2), 355–371. https://doi.org/10.5367/te.2013.0276

Amara, N., Landry, R., & Doloreux, D. (2009). Patterns of innovation in knowledge-intensive business services. *The Service Industries Journal, 29*(4), 407–430. https://doi.org/10.1080/02642060802307847

Amara, N., Landry, R., & Traoré, N. (2008). Managing the protection of innovations in knowledge-intensive business services. *Research Policy, 37*(9), 1530–1547. https://doi.org/10.1016/j.respol.2008.07.001

Antonelli, C. (2017). The derived demand for knowledge. *Economics of Innovation and New Technology, 26*(1–2), 183–194. https://doi.org/10.1080/10438599.2016.1202518

Antonietti, R., Cainelli, G., & Lupi, C. (2013). Vertical disintegration and spatial co-localization: The case of Kibs in the metropolitan region of Milan. *Economics Letters, 118*(2), 360–363. https://doi.org/10.1016/j.econlet.2012.11.031

Asikainen, A. L. (2013). Innovation modes and strategies in knowledge intensive business services. *Service Business*, 1–19. https://doi.org/10.1007/s11628-013-0219-5

Aslesen, H. W., & Isaksen, A. (2007a). Knowledge intensive business services and urban industrial development. *The Service Industries Journal, 27*(3), 321–338.

Aslesen, H. W., & Isaksen, A. (2007b). New perspectives on knowledge-intensive services and innovation. *Geografiska Annaler, 89B*(S1), 45–58.

Aslesen, H. W., Isaksen, A., & Stambol, L. S. (2004). KIBS and industrial development of cities. Labour mobility, innovation and client interaction. In *44th Congress of the European Regional Science Association*, 1–29.

Baláž, V. (2004). Knowledge-intensive business services in transition economies. *The Service Industries Journal, 24*(4), 83–100. https://doi.org/10.1080/0264206042000275208

Basilioa, O., Laredob, P., & Sanchezc, P. (2019). The organization of R&D activities in large knowledge intensive business services: The case of a "Big Four" consultancy. *International Journal of Innovation Management, 23*(3). https://doi.org/10.1142/S1363919619500270

Bettiol, M., De Marchi, V., Di Maria, E., & Grandinetti, R. (2012). Determinants of market extension in knowledge-intensive business services: Evidence from a regional innovation system. *European Planning Studies, 21*(4), 1–18. https://doi.org/10.1080/09654313.2012.722930

Bettiol, M., Di Maria, E., & Grandinetti, R. (2011). Market extension and knowledge management strategies of knowledge-intensive business services. *Knowledge Management Research & Practice, 9*(4), 305–314. https://doi.org/10.1057/kmrp.2011.35

Bocquet, R., Brion, S., & Mothe, C. (2016). The role of cluster intermediaries for KIBS' resources and innovation. *Journal of Small Business Management, 54*, 256–277. https://doi.org/10.1111/jsbm.12298

Bolisani, E., Donò, A., & Scarso, E. (2016a). Relational marketing in knowledge-intensive business services: An analysis of the computer services sector. *Knowledge Management Research and Practice, 14*(3), 319–328. https://doi.org/10.1057/kmrp.2014.35

Bolisani, E., Paiola, M., & Scarso, E. (2013). Knowledge protection in knowledge-intensive business services. *Journal of Intellectual Capital, 14(2)*, 192–211. https://doi.org/10.1108/14691931311323841

Bolisani, E., Scarso, E., & Giuman, L. (2016b). Knowledge management in client-supplier relationship: Emergent vs deliberate approach in small KIBS. *Knowledge Management Research and Practice, 14*(2), 178–185. https://doi.org/10.1057/kmrp.2015.11

Bolisani, E., Scarso, E., & Zieba, M. (2016c). How to deal with knowledge in small companies? Defining emergent Km approach. *International Journal of Learning and Intellectual Capital, 13* (2–3), 104–118. https://doi.org/10.1504/IJLIC.2016.075701

Borodako, K., Berbeka, J., & Rudnicki, M. (2014). The potential of local KIBS companies as a determinant of tourism development in Krakow. *Tourism Economics, 20*(6), 1337–1348. https://doi.org/10.5367/te.2013.0351

Borodako, K., Berbeka, J., & Rudnicki, M. (2015). Tourism enterprises as buyers of knowledge-intensive business services. *SAGE Open, 5*, 1–10. https://doi.org/10.1177/2158244015569669

Braga, A., Marques, C. S., & Serrasqueiro, Z. (2018). Internationalisation strategy of knowledge-intensive business services. *Journal of the Knowledge Economy, 9*(2), 359–377. https://doi.org/10.1007/s13132-017-0461-5

Braga, A. M., Marques, C. S., & Serrasqueiro, Z. M. (2017). KIBS' key dimensions: A qualitative study on innovation, knowledge, networks, location and internationalisation. *International Journal of Entrepreneurship and Small Business, 32*(3), 364–385. https://doi.org/10.1504/IJESB.2017.087060

Brenner, T., Capasso, M., Duschl, M., Frenken, K., & Treibich, T. (2018). Causal relations between knowledge-intensive business services and regional employment growth. *Regional Studies, 52* (2), 172–183. https://doi.org/10.1080/00343404.2016.1265104

Bujdosó, Z., Pénzes, J., Dávid, L., & Madaras, S. (2016). The spatial pattern of KIBS and their relations with the territorial development in Romania. *Amfiteatru Economic, 18*(41), 73–88.

Bumberová, V., & Milichovský, F. (2019). Sustainability development of knowledge-intensive business services: Strategic actions and business performance. *Sustainability (Switzerland), 11* (18). https://doi.org/10.3390/su11185136

Busu, C., & Busu, M. (2017). The role of knowledge intensive business services on romania's economic revival and modernization at the regional level. *Sustainability (Switzerland), 9*(4). https://doi.org/10.3390/su9040526

Cabigiosu, A., & Campagnolo, D. (2019). Innovation and growth in KIBS: The role of clients' collaboration and service customisation. *Industry and Innovation, 26*(5), 592–618. https://doi.org/10.1080/13662716.2018.1483823

Cabigiosu, A., Campagnolo, D., Furlan, A., & Costa, G. (2015). Modularity in KIBS: the case of third-party logistics service providers. *Industry and Innovation, 22*(2), 126–146. https://doi.org/10.1080/13662716.2015.1023012

Calvo, N., Rodeiro-Pazos, D., & Fernández-López, S. (2017). Science and technology parks as accelerators of knowledge-intensive business services. A case study. *International Journal of Business and Globalisation, 18*(1), 42–57. https://doi.org/10.1504/IJBG.2017.081025

Campisi, D., Mancuso, P., Mastrodonato, S. L., & Morea, D. (2019). Efficiency assessment of knowledge intensive business services industry in Italy: Data envelopment analysis (DEA) and financial ratio analysis. *Measuring Business Excellence, 23*(4), 484–495. https://doi.org/10.1108/MBE-09-2019-0095

Chichkanov, N., Miles, I., & Belousova, V. (2019). Drivers for innovation in KIBS: Evidence from Russia. *Service Industries Journal*. https://doi.org/10.1080/02642069.2019.1570151

Chung, S. (2002). Building a national innovation system through regional innovation systems. *Technovation, 22*(8), 485–491. https://doi.org/10.1016/S0166-4972(01)00035-9

Chung, T.-A., & Tseng, C.-Y. (2019). The knowledge intensity and the economic performance in Taiwan's knowledge intensity business services. *Economic Research-Ekonomska Istrazivanja, 32*(1), 797–811. https://doi.org/10.1080/1331677X.2019.1583586

Ciriaci, D., Montresor, S., & Palma, D. (2015). Do KIBS make manufacturing more innovative? An empirical investigation of four European countries. *Technological Forecasting and Social Change, 95*, 135–151. https://doi.org/10.1016/j.techfore.2015.02.008

Consoli, D., & Elche-Hortelano, D. (2010). Variety in the knowledge base of knowledge intensive business services. *Research Policy, 39*(10), 1303–1310. https://doi.org/10.1016/j.respol.2010.08.005

Corrocher, N., & Cusmano, L. (2014). The 'KIBS Engine' of regional innovation systems: Empirical evidence from European regions. *Regional Studies, 48*(7), 1212–1226. https://doi.org/10.1080/00343404.2012.731045

Corrocher, N., Cusmano, L., & Lenzi, C. (2013). Growth in knowledge-intensive business services: evidence from Lombardy. *Industry and Innovation, 20*(6), 563–584. https://doi.org/10.1080/13662716.2013.833376

Corrocher, N., Cusmano, L., & Morrison, A. (2009). Modes of innovation in knowledge-intensive business services evidence from Lombardy. *Journal of Evolutionary Economics, 19*, 173–196. https://doi.org/10.1007/s00191-008-0128-2

Corsi, C., Prencipe, A., Rodríguez-Gulías, M. J., Rodeiro-Pazos, D., & Fernández-López, S. (2019). Growth of KIBS and non-KIBS firms: Evidences from university spin-offs | 知识密集型服务业和非知识密集型服务业企业的增长:来自大学衍生企业的证据. *Service Industries Journal, 39*(1), 43–64. https://doi.org/10.1080/02642069.2018.1436703

Czarnitzki, D., & Spielkamp, A. (2003). Business services in Germany: Bridges for innovation. *The Service Industries Journal, 23*(2), 1–30. https://doi.org/10.1080/02642060412331300862

den Hertog, P. (2000). Knowledge-intensive business services as co-producers of innovation. *International Journal of Innovation Management, 4*(4), 491–528.

Desmarchelier, B., Djellal, F., & Gallouj, F. (2013). Knowledge intensive business services and long term growth. *Structural Change and Economic Dynamics, 25*(1), 188–205. https://doi.org/10.1016/j.strueco.2012.07.003

Desyllas, P., Miozzo, M., Lee, H.-F., & Miles, I. (2018). Capturing value from innovation in knowledge-intensive business service firms: The role of competitive strategy. *British Journal of Management, 29*(4), 769–795. https://doi.org/10.1111/1467-8551.12273

Deza, X. V., & Lopez, M. G. (2014). Regional concentration of knowledge-intensive business services in Europe. *Environment and Planning C: Government and Policy, 32*(6), 1036–1058. https://doi.org/10.1068/c11171r

Doloreux, D., Freel, M., & Shearmur, R. (Eds.). (2010). *Knowledge-intensive business services. Geography and innovation*. Ashgate.

Doloreux, D., Shearmur, R., & Rodriguez, M. (2018). Internal R&D and external information in knowledge-intensive business service innovation: Complements, substitutes or independent? *Technological and Economic Development of Economy, 24*(6), 2255–2276. https://doi.org/10.3846/tede.2018.5694

Doloreux, D., Shearmur, R., & Van Assche, A. (2019a). Combined innovation and export strategies of KIBS in different regional settings. *Industry and Innovation, 26*(6), 715–740. https://doi.org/10.1080/13662716.2018.1498769

Doloreux, D., Turkina, E., & Van Assche, A. (2019b). Innovation type and external knowledge search strategies in KIBS: Evidence from Canada. *Service Business, 13*(3), 509–530. https://doi.org/10.1007/s11628-018-00393-y

Fernandes, C. I., & Ferreira, J. J. M. (2013). Knowledge spillovers: Cooperation between universities and KIBS. *R and D Management, 43*(5), 461–472. https://doi.org/10.1111/radm.12024

Figueiredo, R., & de Matos Ferreira, J. J. (2019). Spinner model: Prediction of propensity to innovate based on knowledge-intensive business services. *Journal of the Knowledge Economy*. https://doi.org/10.1007/s13132-019-00607-2

Figueiredo, R., Ferreira, J. J. M., Silveira, R. G., & Villarinho, A. T. (2020). Innovation and co-creation in knowledge intensive business services: the Spinner model. *Business Process Management Journal, 26*(4, SI), 909–923. https://doi.org/10.1108/BPMJ-10-2019-0424

Figueiredo, R., Quelhas, O., Neto, J. V., & Ferreira, J. J. (2019). The role of knowledge intensive business services in economic development: A bibliometric analysis from Bradford, Lotka and Zipf laws. *Gestao e Producao, 26*(4). https://doi.org/10.1590/0104-530X4356-19

Figueroa-Armijos, M. (2019). Does public entrepreneurial financing contribute to territorial servitization in manufacturing and KIBS in the United States? *Regional Studies, 53*(3), 341–355. https://doi.org/10.1080/00343404.2018.1554900

Freel, M. (2006). Patterns of technological innovation in knowledge-intensive business services. *Industry & Innovation, 13*(3), 335–358. https://doi.org/10.1080/13662710600859157

Gottschalk, P. (2005). *Strategic knowledge management technology*. Macmillan.

Growe, A. (2019). Developing trust in face-to-face interaction of knowledge-intensive business services (KIBS). *Regional Studies, 53*(5), 720–730. https://doi.org/10.1080/00343404.2018.1473567

Hauknes, J., & Antonelli, C. (1999). Knowledge intensive services - What is their role? In *Paper for the OECD Business and Industry Policy Forum, STEP Group* (Issue September). STEP Group.

Heikka, E.-L., Frandsen, T., & Hsuan, J. (2018). Matching value propositions with varied customer needs: The role of service modularity. *Knowledge and Process Management, 25*(1), 64–73. https://doi.org/10.1002/kpm.1556

Heikka, E.-L., & Mustak, M. (2017). Purchasing of knowledge-intensive business services: A case study of relevant factors. *International Journal of Procurement Management, 10*(1), 21–37. https://doi.org/10.1504/IJPM.2017.080909

Heikka, E.-L., & Nätti, S. (2018). Evolving value propositions in knowledge-intensive business services. *Journal of Business and Industrial Marketing, 33*(8), 1153–1164. https://doi.org/10.1108/JBIM-12-2017-0306

Heim, I., Han, T., & Ghobadian, A. (2018). Value co-creation in ICT services company: A case study of a cross-border acquisition. *Journal of East-West Business, 24*(4), 319–338. https://doi.org/10.1080/10669868.2018.1467841

Hu, T. S. (2017). Developments in interactive relationships and knowledge between KIBS firms and their clients in Taiwan. *Knowledge Management Research and Practice, 15*(2), 257–271. https://doi.org/10.1057/s41275-016-0046-2

Hu, T. T.-S., Yu, C. C.-W., Chia, P.-C. P., & Del Giudice, M. (2018). Knowledge exchange types and strategies on the innovation interactions between KIBS firms and their clients in Taiwan. *Cogent Business & Management, 5*(1), 1–23. https://doi.org/10.1080/23311975.2018.1534527

Huggins, R. (2011). The growth of knowledge-intensive business services: Innovation, markets and networks. *European Planning Studies, 19*(8), 1459–1480. https://doi.org/10.1080/09654313.2011.586172

Javalgi, R.(. R.). G., Gross, A. C., Joseph, W. B., & Granot, E. (2011). Assessing competitive advantage of emerging markets in knowledge intensive business services. *Journal of Business & Industrial Marketing, 26*(3), 171–180. https://doi.org/10.1108/08858621111115895

Jesson, J. K., Matheson, L., & Lacey, F. M. (2011). *Doing your literature review. Traditional and systematic techniques*. SAGE.

Johnston, A., & Huggins, R. (2017). University-industry links and the determinants of their spatial scope: A study of the knowledge intensive business services sector. *Papers in Regional Science, 96*(2), 247–260. https://doi.org/10.1111/pirs.12185

Kamp, B., & Ruiz de Apodaca, I. (2017). Are KIBS beneficial to international business performance: Evidence from the Basque Country. *Competitiveness Review, 27*(1), 80–95. https://doi.org/10.1108/CR-07-2015-0066

Koch, A., & Stahlecker, T. (2006). Regional innovation systems and the foundation of knowledge intensive business services. A comparative study regional innovation systems and the foundation of knowledge intensive business services. A comparative study in Bremen, Munich, and Stuttgart. *European Planning Studies, 14*, 37–41.

Koch, A., & Strotmann, H. (2008). Absorptive capacity and innovation in the knowledge intensive business service sector. *Economics of Innovation and New Technology, 17*(6), 511–531. https://doi.org/10.1080/10438590701222987

Kuula, S., Haapasalo, H., & Tolonen, A. (2018). Cost-efficient co-creation of knowledge intensive business services. *Service Business, 12*(4), 779–808. https://doi.org/10.1007/s11628-018-0380-y

Lafuente, E., Vaillant, Y., & Leiva, J. C. (2018). Sustainable and traditional product innovation without scale and experience, but only for KIBS! *Sustainability (Switzerland), 10*(4). https://doi.org/10.3390/su10041169

Lafuente, E., Vaillant, Y., & Vendrell-Herrero, F. (2017). Territorial servitization: Exploring the virtuous circle connecting knowledge-intensive services and new manufacturing businesses. *International Journal of Production Economics, 192*, 19–28. https://doi.org/10.1016/j.ijpe.2016.12.006

Lafuente, E., Vaillant, Y., & Vendrell-Herrero, F. (2019). Territorial servitization and the manufacturing renaissance in knowledge-based economies. *Regional Studies, 53*(3), 313–319. http://10.0.4.56/00343404.2018.1542670

Landry, R., Amara, N., & Doloreux, D. (2012). Knowledge-exchange strategies between KIBS firms and their clients. *The Service Industries Journal, 32*, 291–320. https://doi.org/10.1080/02642069.2010.529131

Lara, F., Palacios-Marques, D., & Devece, C. A. (2012). How to improve organisational results through knowledge management in knowledge-intensive business services. *Service Industries Journal, 32*(11), 1853–1863. https://doi.org/10.1080/02642069.2011.574283

Larsen, J. N. (2000). Supplier-user interaction in knowledge-intensive business services: Types of expertise and modes of organization. In I. M. Boden & I. Miles (Eds.), *Services and the knowledge-based economy* (pp. 146–154). Continuum.

Lee, H.-F., & Miozzo, M. (2019). Which types of knowledge-intensive business services firms collaborate with universities for innovation? *Research Policy, 48*(7), 1633–1646. https://doi.org/10.1016/j.respol.2019.03.014

Mangiarotti, G. (2012). Knowledge management practices and innovation propensity: A firm-level analysis for Luxembourg. *International Journal of Technology Management, 58*, 261. https://doi.org/10.1504/IJTM.2012.046618

Miles, I. (1999). *Services in national innovation systems: From traditional services to knowledge intensive business services. Transformation Towards a Learning Economy*. The Finnish National Fund for Research and Development–Sitra. Report, 213.

Miles, I. (2005). Knowledge intensive business services: Prospects and policies. *Foresight, 7*(6), 39–63. https://doi.org/10.1108/14636680510630939

Miles, I. (2008). Patterns of innovation in service industries. *IBM Systems Journal, 47*(1), 115–128. https://doi.org/10.1147/sj.471.0115

Miles, I., Kastrinos, N., Flanagan, K., Bilderbeek, R., Den Hertog, P., Huntink, W., & Bouman, M. (1995). Knowledge-intensive business services. Users, carriers and sources of innovation. *A Report to DG13 SPRINT-EIMS, March*, 1–117.

Miozzo, M., Desyllas, P., Lee, H. F., & Miles, I. (2016). Innovation collaboration and appropriability by knowledge-intensive business services firms. *Research Policy, 45*(7), 1337–1351. https://doi.org/10.1016/j.respol.2016.03.018

Miozzo, M., & Grimshaw, D. (2005). Modularity and innovation in knowledge-intensive business services: IT outsourcing in Germany and the UK. *Research Policy, 34*(9), 1419–1439. https://doi.org/10.1016/j.respol.2005.06.005

Mol, M. J., & Brandl, K. (2018). Bridging what we know: The effect of cognitive distance on knowledge-intensive business services produced offshore. *International Business Review, 27* (3), 669–677. https://doi.org/10.1016/j.ibusrev.2017.11.003

Muller, E., & Doloreux, D. (2009). What we should know about knowledge-intensive business services. *Technology in Society, 31*(1), 64–72. https://doi.org/10.1016/j.techsoc.2008.10.001

Muller, E., & Zenker, A. (2001). Business services as actors of knowledge transformation: The role of KIBS in regional and national innovation systems. *Research Policy, 30*(9), 1501–1516. https://doi.org/10.1016/S0048-7333(01)00164-0

Müller, P., & Franz, M. (2019). Transnational labour migration and the offshoring of knowledge-intensive business services within global production networks: The case of a German automotive company in Turkey. *Environment and Planning A, 51*(6), 1350–1369. https://doi.org/10.1177/0308518X18811353

Musolesi, A., & Huiban, J.-P. (2010). Innovation and productivity in knowledge intensive business services. *Journal of Productivity Analysis, 34*, 63–81. https://doi.org/10.1007/s11123-009-0163-5

Mustak, M. (2019). Customer participation in knowledge intensive business services: Perceived value outcomes from a dyadic perspective. *Industrial Marketing Management, 78*, 76–87. https://doi.org/10.1016/j.indmarman.2017.09.017

O'Farrell, P. N., & Moffat, L. A. R. (1995). Business services and their impact upon client performance: An exploratory interregional analysis. *Regional Studies, 29*, 111–124. https://doi.org/10.1080/00343409512331348843

Orłowska, A., Mazur, Z., & Łaguna, M. (2017). Systematyczny przegląd literatury: Na czym polega i czym się różni od innych przeglądów. *Ogrody Nauk i Sztuk, 7*, 350–363. https://doi.org/10.15503/onis2017.350.363

Päällysaho, S., & Kuusisto, J. (2008). Intellectual property protection as a key driver of service innovation: An analysis of innovative KIBS businesses in Finland and the UK. *International Journal of Services Technology and Management, 9*(3/4), 268. https://doi.org/10.1504/IJSTM.2008.019707

Päällysaho, S., & Kuusisto, J. (2011). Informal ways to protect intellectual property (IP) in KIBS businesses. *Innovation: Management, Policy & Practice, 13*, 62–76. https://doi.org/10.5172/impp.2011.13.1.62

Pace, L. A., & Miles, I. (2019). The influence of KIBS-client interactions on absorptive capacity-building for environmental innovation. *European Journal of Innovation Management, 23*(4), 553–580. https://doi.org/10.1108/EJIM-01-2019-0026

Paiola, M., Bolisani, E., & Scarso, E. (2013). Characterisation of knowledge-based networking strategies: Insights from the KIBS sector. *Internationl Journal of Networking and Virtual Organisations, 13*(3), 263–281.

Palacios-Marques, D., Gil-Pechuán, I., & Lim, S. (2011). Improving human capital through knowledge management practices in knowledge-intensive business services. *Service Business, 5*, 99–112. https://doi.org/10.1007/s11628-011-0104-z

Peiker, W., Pflanz, K., Kujath, H. J., & Kulke, E. (2012). The heterogeneity of internationalisation in knowledge intensive business services. *Zeitschrift Für Wirtschaftsgeographie, 56*(1–2), 209–225.

Petticrew, M., & Roberts, H. (2006). *Systematic reviews in the social sciences. A practical guide.* Blackwell Publishing.

PREST. (2005). *Knowledge-intensive business services: Trends and scenarios* (Issue June). http://www.eurofound.europa.eu/observatories/emcc/articles/working-conditions/knowledge-intensive-business-services-trends-and-scenarios

Qi, H., Liu, S., Qi, W., & Liu, Z. (2019). Geographical concentration of knowledge- and technology-intensive industries and city innovation in China. *Sustainability (Switzerland), 11* (18). https://doi.org/10.3390/su11184840

Rodríguez, A., & Nieto, M. J. (2010). Cooperation and innovation in the internationalisation of knowledge-intensive business services. In *Progress in International Business Research* (Vol. 5). Elsevier. https://doi.org/10.1108/S1745-8862(2010)0000005015

Rodríguez, A., & Nieto, M. J. (2012). The internationalization of knowledge-intensive business services: The effect of collaboration and the mediating role of innovation. *The Service Industries Journal, 32*(7), 1057–1075. https://doi.org/10.1080/02642069.2012.662493

Rodríguez, A., Nieto, M. J., & Santamaría, L. (2018). International collaboration and innovation in professional and technological knowledge-intensive services. *Industry and Innovation, 25*(4), 408–431. https://doi.org/10.1080/13662716.2017.1414752

Rodriguez, M. (2014). Innovation, knowledge spillovers and high-tech services in European regions. *Engineering Economics, 25*(1), 31–39. https://doi.org/10.5755/j01.ee.25.1.3207

Rodriguez, M., Doloreux, D., & Shearmur, R. (2015). Innovation strategies, innovator types and openness: A study of KIBS firms in Spain. *Service Business*. https://doi.org/10.1007/s11628-015-0286-x

Scarso, E. (2015). What do we know about KIBS? Results of a systematic literature review. In J. C. Spender, G. Schiuma, & V. Albino (Eds.), *IFKAD 2015: 10TH international forum on knowledge asset dynamics: Culture, innovation and entrepreneurship: connecting the knowledge dots* (pp. 1159–1172). IKAM-INST Knowledge Asset Management.

Scarso, E., & Bolisani, E. (2012). Trust in knowledge exchanges between service providers and clients: A multiple case study of KIBS. *Knowledge Management Research & Practice, 10*(1), 16–26. https://doi.org/10.1057/kmrp.2011.28

Schmidt, S. (2015). Balancing the spatial localisation "Tilt": Knowledge spillovers in processes of knowledge-intensive services. *Geoforum, 65*, 374–386. https://doi.org/10.1016/j.geoforum.2015.05.009

Shearmur, R., & Doloreux, D. (2019). KIBS as both innovators and knowledge intermediaries in the innovation process: Intermediation as a contingent role. *Papers in Regional Science, 98*(1), 191–209. https://doi.org/10.1111/pirs.12354

Siikonen, J., Pellikka, J., & Littunen, H. (2011). Knowledge networks in innovation among small KIBS firms. *International Journal of Innovation and Regional Development, 3*(5), 515–534.

Skórska, A. (2012). *Wiedzochłonne usługi biznesowe w Polsce i innych krajach Unii Europejskiej.* Wydawnictwo Uniwersytetu Ekonomicznego w Katowicach.

Smedlund, A., & Toivonen, M. (2007). The role of KIBS in the IC development of regional clusters. *Journal of Intellectual Capital, 8*(1), 159–170. https://doi.org/10.1108/14691930710715114

Stahlecker, T., & Koch, A. (2004, April). On the significance of economic structure and regional innovation systems for the foundation of knowledge-intensive business services. Working Paper.

Strambach, S. (2001). Innovation processes and the role of knowledge-intensive business services (KIBS). In K. Koschatzky, M. Kulicke, & A. Zenker (Eds.), *Innovation networks: Concepts and challenges in the European perspective* (pp. 53–68). Physica-Verlag HD. https://doi.org/10.1007/978-3-642-57610-2_4

Teirlinck, P. (2018). Pathways for knowledge exchange in SMEs in software-driven knowledge-intensive business services. *R and D Management, 48*(3), 343–353. https://doi.org/10.1111/radm.12311

The European Cluster Observatory. (2009). *Priority sector report: Knowledge intensive business services* (Issue March). www.clusterobservatory.eu/common/galleries/downloads/kibs.pdf

Thomi, W., & Böhn, T. (2003). Knowledge intensive business serevices in regional systems of innovation - The case of Southeast-Finland. In *43rd European Congress of the Regional Science Association*.

Toivonen, M. (2004). Expertise as business. Long-term development and future prospects of knowledge-intensive business services (KIBS) [Helsinki University of Technology]. In *Doctoral dissertation series* (Vol. 2). http://lib.tkk.fi/Diss/2004/isbn9512273152/

Toivonen, M. (2007). Innovation policy in services: The development of knowledge-intensive business services (KIBS) in Finland. *Innovation: Management, Policy & Practice, 9*, 249–261. https://doi.org/10.5172/impp.2007.9.3-4.249

Toivonen, M., Brax, S., & Tuominen, T. (2008). Client-oriented multicompetence: The core asset in KIBS. *International Journal of Services Technology and Management, 10*(2/3/4), 175. https://doi.org/10.1504/IJSTM.2008.022118

Tranfield, D., Denyer, D., & Smart, P. (2003). Towards a methodology for developing evidence-informed management knowledge by means of systematic review. *British Journal of Management, 14*(3), 207–222.

Tseng, C.-Y., Pai, D. C., & Hung, C.-H. (2011). Knowledge absorptive capacity and innovation performance in KIBS. *Journal of Knowledge Management, 15*(6), 971–983. https://doi.org/10.1108/13673271111179316

Tuominen, T., & Toivonen, M. (2011). Studying innovation and change activities in KIBS through the lens of innovative behaviour. *International Journal of Innovation Management, 15*(2), 393–422. https://doi.org/10.1142/S1363919611003209

Viljamaa, A., Kolehmainen, J., & Kuusisto, J. (2010). For and against? An exploration of inadvertent influences of policies on KIBS industries in the Finnish policy setting. *The Service Industries Journal, 30*, 71–84. https://doi.org/10.1080/02642060802238513

Werner, P., & Strambach, S. (2018). Policy mobilities, territorial knowledge dynamics and the role of KIBS: Exploring conceptual synergies of formerly discrete approaches. *Geoforum, 89*, 19–28. https://doi.org/10.1016/j.geoforum.2017.12.007

Windrum, P., & Tomlinson, M. (1999). Knowledge-intensive services and international competitiveness: A four country comparison. *Technology Analysis & Strategic Management, 11*(3), 391–408. https://doi.org/10.1080/095373299107429

Wojtczuk-Turek, A. (2017). In search of key HR practices for improvement of productivity of employees in the KIBS sector. *Contemporary Economics, 11*(1), 5–16. https://doi.org/10.5709/ce.1897-9254.225

Wong, P. K., & He, Z.-L. (2005). A comparative study of innovation behaviour in Singapore's KIBS and manufacturing firms. *The Service Industries Journal, 25*, 23–42. https://doi.org/10.1080/0264206042000302397

Wood, P. (2002). Knowledge-intensive services and urban. *Urban Studies, 39*(5–6), 993–1002. https://doi.org/10.1080/00420980220128 41

Wood, P. (2006). The regional significance of knowledge-intensive services in Europe. *Innovation: The European Journal of Social Science Research, 19*(1), 51–66. https://doi.org/10.1080/13511610600607718

Wood, P. A., Bryson, J., & Keeble, D. (1993). Regional patterns of small firm development in the business services - Evidence from the United-Kingdom. *Environment and Planning A, 25*, 677–700. https://doi.org/10.1068/a250677

Wyszkowska-Kuna, J. (2016). *Usługi biznesowe oparte na wiedzy. Wpływ na konkurencyjność gospodarki na przykładzie wybranych krajów Unii Europejskiej*. Wydawnictwo Uniwersytetu Łódzkiego.

Wyszkowska-Kuna, J. (2017a). The role of intermediate demand and technology for international competitiveness of the KIBS sector: Evidence from European Union countries. *Journal of International Trade and Economic Development, 26*(7), 777–800. https://doi.org/10.1080/09638199.2017.1302495

Wyszkowska-Kuna, J. (2017b). Knowledge-intensive business services and the innovativeness of manufacturing companies in the European Union. *Gospodarka Narodowa, 4*, 99–120.

Xin, Y., Chai, K.-H., Ojanen, V., & Brombacher, A. (2013). The moderating effects of service solution characteristics on competitive advantage. *Service Industries Journal, 33*(15–16), 1633–1658. https://doi.org/10.1080/02642069.2011.638918

Yam, R. C. M. M., Lo, W., Tang, E. P. Y. Y., & Lau, A. K. W. W. (2011). Analysis of sources of innovation, technological innovation capabilities, and performance: An empirical study of Hong Kong manufacturing industries. *Research Policy, 40*(3), 391–402. https://doi.org/10.1016/j.respol.2010.10.013

Yum, S. (2019). The interaction between knowledge-intensive business services and urban economy. *Annals of Regional Science, 63*(1), 53–83. https://doi.org/10.1007/s00168-019-00920-3

Zenka, J., Novotny, J., Slach, O., Ivan, I., Ženka, J., Novotný, J., Slach, O., & Ivan, I. (2017). Spatial distribution of knowledge-intensive business services in a small post-Communist economy. *Journal of the Knowledge Economy, 8*(2), 385–406. https://doi.org/10.1007/s13132-015-0260-9

Zieba, M. (2013). Knowledge-intensive business services (KIBS) and their role in the knowledge-based economy. In *Proceedings of the European Conference on Knowledge Management, ECKM, 2*, 785–792.

Zieba, M., Bolisani, E., Paiola, M., & Scarso, E. (2017). Searching for innovation knowledge: Insight into KIBS companies. *Knowledge Management Research and Practice, 15*(2), 282–293. https://doi.org/10.1057/s41275-017-0050-1

Zieba, M., Bolisani, E., & Scarso, E. (2016). Emergent approach to knowledge management by small companies: Multiple case-study research. *Journal of Knowledge Management, 20*(2), 292–307. https://doi.org/10.1108/JKM-07-2015-0271

Zieba, M., & Konczynski, P. (2019). Factors of successful client co-production in knowledge-intensive business services case study analysis. *Kybernetes, 49*(1), 141–164. https://doi.org/10.1108/K-04-2019-0297

Zieba, M., & Zieba, K. (2014). Knowledge management critical success factors and the innovativeness of KIBS companies. *Inzinerine Ekonomika-Engineering Economics, 25*(4), 458–465. https://doi.org/10.5755/j01.ee.25.4.6494

5 Knowledge Flows in KIBS Companies: Theoretical Framework and Study Description

The theoretical framework of the research is based on the broadly understood theory of knowledge management and the concept of knowledge flows. The starting point is theoretical considerations presented in the previous parts of the book, concerning knowledge, knowledge management, knowledge transfer and knowledge flows. These considerations will be presented in the following parts of this section in the context of companies offering knowledge-intensive business services and their specificity. On their basis, research questions were formulated that formed the basis of the empirical research.

5.1 Knowledge and Approaches to Knowledge Management in Companies Offering KIBS

In addition to the universal features of knowledge presented in one of the previous sections, the following aspects are related to knowledge in KIBS companies:

- *Difficulty in evaluating the value of knowledge, both in terms of inputs and outputs*—Due to the elusiveness of knowledge, knowledge is difficult to evaluate. Because in KIBS companies, knowledge is both an input and a result of actions (Desmarchelier et al., 2013; Tuominen & Toivonen, 2011), valuation of the KIBS value can be problematic.
- *Fluency of knowledge*—the risk of knowledge loss by companies from the KIBS sector is very important for the functioning of these organizations (Zięba, 2017).
- *Difficulty in securing the rights to knowledge by the organization*—there is a risk that KIBS will lose its competitive advantage if its key knowledge is acquired by competitors (Amara et al., 2008; Päällysaho & Kuusisto, 2011).
- *Risk of knowledge obsolescence and the need to constantly refresh it*—without access to up-to-date knowledge, KIBS cannot provide its services to clients. As stated by Pardos et al. (2007), KIBS companies provide technologies, new

M. Zieba, *Understanding Knowledge-Intensive Business Services*, Knowledge Management and Organizational Learning 10,
https://doi.org/10.1007/978-3-030-75618-5_5

production/sales techniques and new management methods, and therefore their knowledge must be up-to-date and include the latest achievements in a given field.

- *Responsibility for the inappropriate use of knowledge or the provision of inappropriate knowledge to companies—customers* (a characteristic feature of KIBS companies)—it is the responsibility of the KIBS companies that offer services on which the company's situation—the customer depends, to convey the right knowledge in an appropriate way (so that the customer is able to understand and assimilate it). For example, if KIBS offers legal services, misrepresentation of knowledge or providing clients with incorrect knowledge may result in serious legal consequences (e.g. trial, penalty, etc.).
- *The result of the process of combining the knowledge possessed by KIBS with the knowledge of the client in the form of the service provided is very often uncertain and difficult to predict*—the cooperation between the client and KIBS is influenced by several factors, e.g. the ability to learn in the client–supplier relationship (Kohtamäki & Partanen, 2016) or trust, teamwork and communication (Zieba & Kończyński, 2017). If any of these elements is missing, the result of the cooperation between KIBS and the customer may not be as expected by either party. It is also difficult to predict what the final effect of this cooperation will look like.

As for the types of knowledge that may be relevant to the activities of KIBS companies, they can be classified according to the knowledge possessed by different entities and knowledge of various issues. The knowledge possessed by various entities (inside and outside the organization) includes knowledge of employees and owner, knowledge of customers, knowledge of suppliers, knowledge of cooperators, knowledge of research and development units (Zieba et al., 2017). In turn, knowledge of various topics includes:

- knowledge of the market and customers;
- knowledge about the product/service produced by the company;
- knowledge of key business processes;
- knowledge about management—especially the knowledge on the basis of which key decisions are made;
- knowledge of human resources—the skills and experience of the organization's employees;
- knowledge of knowledge gaps,[1] which the organization needs to complete;
- knowledge of suppliers and partners;
- knowledge about trends.

[1]The knowledge gap indicates a mismatch between what the organization needs to know and what it actually knows (Perrott, 2007). For example, Dalkir (2011, p. 325) proposed a set of questions to identify missing knowledge in the organization, such as: "What kind of questions do you keep asking and getting no answers to?" or "How much time do you spend looking for the necessary knowledge?".

The knowledge possessed by KIBS companies, like the knowledge in other companies, may take various forms, e.g. codified and non-codified, overt and hidden. On the one hand, companies providing KIBS services base their activities on codified knowledge and transferred to the client, on the other hand, the tacit knowledge possessed by employees of KIBS companies is usually necessary to prepare and deliver the KIBS service. At KIBS, the relationship between explicit and tacit knowledge and the transformation of these types of knowledge are also important. Tacit knowledge possessed by employees, which is key knowledge for KIBS, should be transferred into explicit knowledge, available to other employees, as well as to customers who need it to perform their tasks.

As KIBS provides its clients with a knowledge-intensive service in a selected area, it usually needs to have some specialist knowledge. It can be technological knowledge (e.g. in the case of an IT company offering IT systems) or non-technological knowledge (e.g. in the case of a company providing accounting services). Some KIBS firms rely on both types of expertise to provide different types of services (both technology-based and non-technology based).

In the case of KIBS companies, the importance of knowledge as a strategic resource is very high. The company's knowledge base consists of the contribution of knowledge and information, know-how, as well as the abilities and skills of members of this type of organization. The specific knowledge of the KIBS company, which is an important "isolating mechanism" from competitors, results from the constant search and accumulation of new solutions, based and grounded in the created knowledge base of the organization. KIBS companies run "activities where knowledge is both the main input and the main output"(Gallouj, 2002). Therefore, they should pay particular attention to the possessed knowledge resources and the possibility of acquiring new ones. In the case of KIBS companies, in order for knowledge to be a strategic resource, it should have the following characteristics:

- *it should be potentially important/significant, so that companies–customers want to use it*—in other words, the knowledge possessed by KIBS should also be the knowledge missing in the company—the customer;
- *should be unique/exceptional*—the uniqueness of KIBS's knowledge lies in the fact that it combines not only knowledge in a given area, but also knowledge from previously developed and delivered products/services;
- *it should be constantly developed by KIBS*—knowledge development involves not only the improvement of products/services for customers, but also its updating and replacement of obsolete knowledge/solutions with new/new ones;
- *it should be difficult to copy/imitate*—in the case of KIBS companies, a large part of their knowledge is the property of their employees—they combine their knowledge with the knowledge possessed by KIBS and the knowledge obtained from and about the client company—such a combination makes this knowledge difficult to copying/imitating, provided that the company's employees are retained and not taken over by a competing organization.

To sum up, it can be stated that KIBS companies, as knowledge-intensive companies, should attach great importance to knowledge, both possessed and possible to obtain. However, the following issues have yet to be investigated:

- *What knowledge is important for the functioning of these companies?*
- *What knowledge do companies in this sector perceive as the most important for their activities?*
- *How can knowledge affect the competitive advantage of the surveyed companies?*

These questions will be answered on the basis of the qualitative research presented in the next chapters.

When analyzing the approaches to knowledge management and their application in KIBS companies, it should be noted that in companies of this type there is a development of the classical perception of knowledge conversion processes, because in traditional terms they concern the inside of the organization and take place in the organizational context. In the case of companies from the KIBS sector, this classic approach should be supplemented with interactions with customers and other potential partners (e.g. universities, R&D units, etc.). When creating the KIBS service, the SECI model (Nonaka et al., 2000) must take place not only in the KIBS company, but also in the client company and, at the same time, between KIBS representatives and the client company representatives. In the case of companies from the KIBS sector, the statement "if only we knew what our clients know" is of particular importance (Gibbert et al., 2002), being a paraphrase of a former HP CEO: "If only we knew what we know".

Interactions between tacit and explicit knowledge on the part of both KIBS and its clients will tend to increase in scale and pace faster if more employees of these organizations are involved. In other words, the creation of organizational knowledge can be seen as a knowledge spiral, starting from the individual level, through the collective (group) level and organizational level, up to the inter-organizational level (Nonaka, 1994).

In the case of the resource approach to KM (Leonard-Barton, 1995), the model of "knowledge sources" fits well in the context of KIBS companies for several reasons. First of all, these companies are largely based on the knowledge and skills of their employees—it is the knowledge and experience of employees that underpin the key skills of this type of organization. Secondly, KIBS companies continuously import knowledge from their environment, especially from their customers and partners. Without customers' knowledge, the provision of KIBS services would often not be possible. Third, KIBS companies implement and integrate their knowledge, creating their services and delivering them to clients. The experience gained during cooperation with other clients allows for a better understanding of future clients and more accurate meeting of their expectations. Collaborative problem solving is something natural in KIBS companies, as teams usually work to create and deliver KIBS services, overcoming problems and doubts. Experimentation is also a natural thing for KIBS companies that try to tailor their services to the needs of individual clients and look to the future, trying to find effective models of cooperation with clients.

The process approach to KM (Bukowitz & Williams, 2000; Davenport & Prusak, 1998; Jashapara, 2011; Murray & Myers, 1997; Probst et al., 2000) can also be successfully applied in KIBS companies, because these companies intensively acquire, create, store, organize, apply, disseminate and use knowledge as part of their daily operations. It should even be expected that these processes will be much more dynamic in KIBS companies than in other types of companies.

Companies from the KIBS sector not only gather knowledge about their current and potential clients (e.g. their needs, motivations, behaviour), but also acquire knowledge from them. Therefore, we can risk a statement that these companies use the concept of CKM (customer knowledge management) (Paquette, 2011) on a daily basis, because they cooperate with their clients in creating services, communicate with them in various ways (both traditional and using new media) and acquire knowledge from their clients on not only their needs in terms of the services offered, but also the organizational specifics and the potential problem in which the KIBS company is supposed to help. Clients can provide KIBS with some unique knowledge that will allow it to learn and improve its internal functioning. These activities often contribute to an increase in the value of services offered, an increase in the level of customer satisfaction, as well as the emergence of service innovations. It can be said that as a rematch, customers receive from KIBS not only knowledge about the products or services offered, helping them to better use them, but also a high level of matching products/services to the actual customer needs. In this way, a strong relationship is created between the customer and KIBS, based on knowledge flows in both directions. The concept of customer knowledge management in the case of KIBS means that it is not enough to simply acquire knowledge from the customer, but also need to be able to combine it with the knowledge and experience of KIBS, creating something new that the customer needs and will be happy with.

To sum up, KIBS companies depend on knowledge not only possessed inside the organization, but also available outside its borders, e.g. in the minds or organizations of clients. The difference between KIBS companies and other organizations is also that in the case of the approaches described above, knowledge is by definition used, e.g., to make decisions by organizations, and in the case of KIBS companies, knowledge is also the main resource in the process of creating and delivering a KIBS type service.

Additionally, due to the high dynamics of new and existing knowledge, organizations should have a different approach to their knowledge and ways of managing it. Gathering knowledge in databases or focusing on a different, static way of collecting knowledge may not be enough. Different types of knowledge are needed in different places of the organization and thanks to the knowledge flows it can reach these places. The originality of KIBS companies is based on the fact that the service they provide is often co-produced in close cooperation with the client and for the needs of a specific client. If this is taken into account, an important element of the KM in KIBS companies will be various types of knowledge flows, both inside the KIBS organization and between the organization and its external environment (including customers, suppliers, partners, R&D centres, cooperators, etc.). These flows will be discussed in the next subchapter.

5.1.1 Knowledge Flows and Their Types in Companies Offering Knowledge-Intensive Business Services

The activity of KIBS companies, due to the repeatedly emphasized dependence on knowledge, is to a large extent related to knowledge flows. In the case of manufacturing companies, we are talking about various types of flows in production processes. The company's resources move through the production process, the purpose of which is to meet customer expectations. Manufacturing companies try to optimize this process, trying to maintain the continuity of its implementation, without unnecessary interruptions and delays.

In the case of KIBS companies, the situation is similar, but with the difference that what flows in the process of creating a service is knowledge in various forms, as well as information and data. The difficulty with knowledge flows is that they often involve tacit knowledge that does not have physical form/form, such as material resources in a traditional manufacturing process. Thus, managing knowledge flows is not an easy task (an employee may, for example, pretend that he has no knowledge, fearing to lose his position in the organization).

Knowledge flows in these types of companies can be divided into various categories and subcategories. With regard to the taxonomy of knowledge flows proposed in the previous section, the first category may be internal flows in the organization, covering its various levels (individual, team, organizational). Internal knowledge flows are important in KIBS companies as they help to spread the specialist knowledge necessary to provide services by these organizations. This knowledge is often hidden in the minds of employees of these companies and, in most cases, cannot be codified and made available in the form of a database. Therefore, it is important to create conditions for the free flow of knowledge in this type of organizations, so that knowledge can be transferred, shared and used for the purpose of performing tasks.

The second category is the flows between KIBS and its environment. These flows include competitors, contractors, universities and R&D units, as well as suppliers and other potential sources of knowledge located outside the boundaries of the organization. Due to the fact that their activities are based on up-to-date, professional knowledge, KIBS companies often use the knowledge of universities and R&D units, where new technologies and solutions are created that are useful in the operation of KIBS companies. Moreover, the sources of knowledge may also be competitors who may become cooperators in some situations.[2] Apart from cooperators, KIBS companies can also gain knowledge from their suppliers.[3] The types of knowledge flows in KIBS are presented in Table 5.1.

[2] An example is the situation when KIBS companies from the same sector unite their forces to jointly complete a large order for a client company. Without the cooperation of organizations competing on a daily basis, the implementation of this contract would not be possible.

[3] More information on the KIBS approach to acquiring knowledge from various internal and external sources for the purpose of creating innovation can be found in the study (Zieba et al., 2017).

Table 5.1 Knowledge flows in KIBS

Knowledge flows		
KIBS organisation[1]	Employee-Employee Employee-Team Employee-Organization Organization-Employee Team-Employee Employee-Supervisor Supervisor-Employee	
Environment and KIBS organization	KIBS–Cooperant[2] Cooperator – KIBS Competitor – KIBS KIBS – Competitor KIBS – University University – KIBS R&D unit - KIBS KIBS – R&D unit	Participation in the service (cooperation/ coopetition/ competition)
	Client (service implementation) Client – KIBS KIBS – Client KIBS-KIBS (internal flows in the KIBS company between the person contacting the client and the rest of the team performing a given KIBS service)	

Source: Own elaboration

In the category of flows between KIBS and the environment, the knowledge flows between the company and customers deserve special attention. The knowledge flowing between KIBS and the client is very often a necessary element to create a KIBS service. Without the flow of knowledge between KIBS and its client, the creation of the KIBS service may be difficult or even impossible. In some cases, the knowledge obtained from the client is the basis for creating the KIBS service. In relation to the client, KIBS participates in various types of knowledge flows, the subject of which may be both explicit and implicit knowledge. This is important for KIBS companies, because, as mentioned before, it is the tacit knowledge possessed by employees that is an important resource of this type of organization. Moreover, when performing the KIBS service, tacit knowledge often has to be articulated in some way, and sometimes simplified and reduced to the form of information and data. In this form, it can be more easily transferred to other team members or representatives of the client company. This data/information or explicit knowledge is then processed by the client, integrated with his knowledge and finally transformed into the knowledge of the client company. In other words, the new knowledge obtained from KIBS is combined with the knowledge existing in the client company—the new knowledge is a missing element, fitting into the existing knowledge structure. Moreover, KIBS companies, especially those offering non-standard services (e.g. innovative technological solutions), often have to educate their clients so that they become aware of the need to purchase a given service

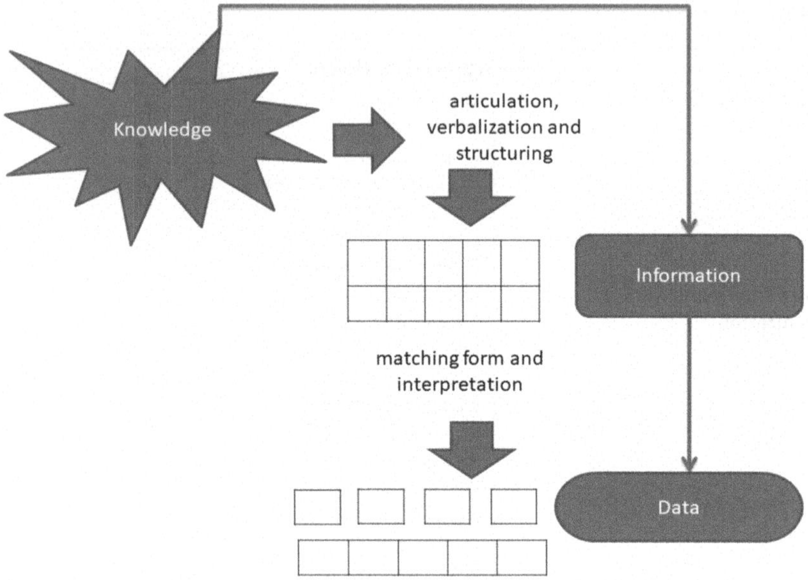

Fig. 5.1 Inverted hierarchy of knowledge–information–data. Source: Author's own illustration, based on: Tuomi (1999)

and be able to use it. In order for the customer to acquire certain knowledge (absorb it), he/she must have a certain knowledge base, without which he/she will not be able to "accept" the service offered by KIBS.

In order for the KIBS company to transfer its knowledge to the client company, it is often necessary to use the so-called inverted hierarchy: knowledge–information–data. In the conventional hierarchy of knowledge, data is at the bottom, which is then converted into information and knowledge. In the inverted hierarchy of knowledge, at the beginning there is knowledge that must be articulated, verbalized and structured to become information, and then—after adjusting form and interpretation—data. These data can then be easily transferred and used (Fig. 5.1).

In addition to the need to simplify the transfer and convert knowledge into information or data, it is often useful to use the so-called the funnel of knowledge. The knowledge funnel is where KIBS consciously limits the amount of knowledge transferred to the client, focusing only on the potentially most important elements (Fig. 5.2).

There are several reasons why KIBS companies limit the transferred knowledge. First, giving the client too many options makes it difficult to choose—the number of solutions to choose from should be limited. Second, the customer often has to be focused on what he/she needs as his/her absorption capacity may be limited (Hamel & Prahalad, 1999, p. 84). In such a situation, transferring too much knowledge could extend the process of acquiring it or even make it impossible. Third, the flow of

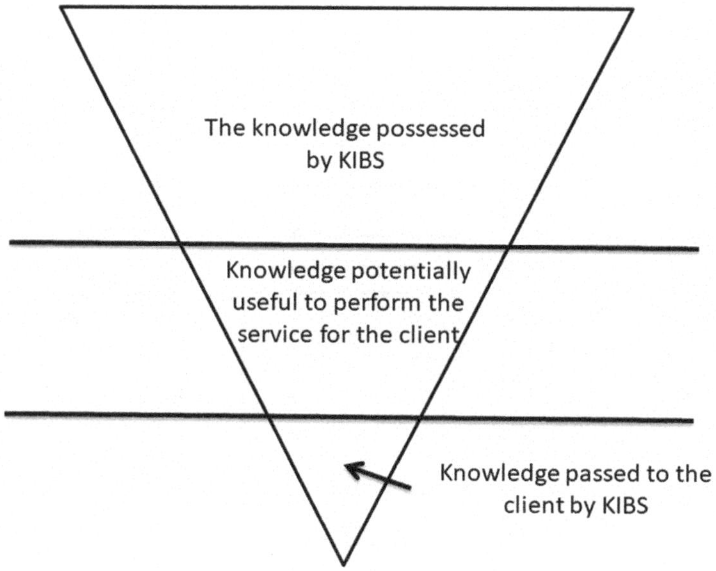

Fig. 5.2 Knowledge funnel used by KIBS companies. Source: Author's own illustration

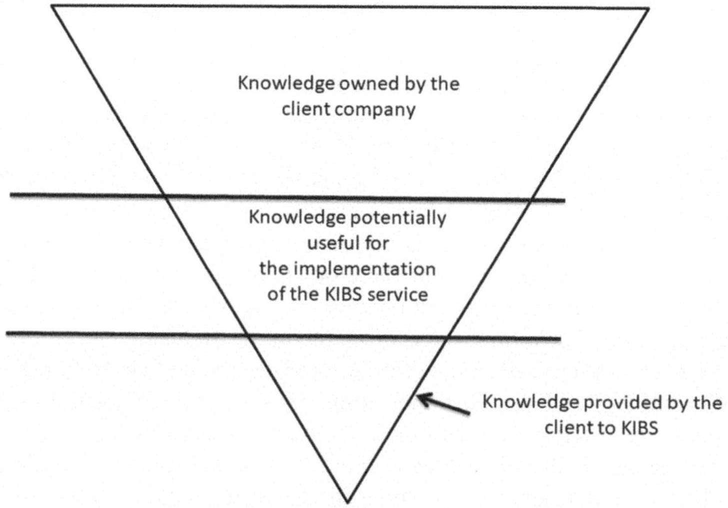

Fig. 5.3 Knowledge funnel used by companies—customers. Source: Author's own illustration

knowledge takes time, so transferring more knowledge will require more time for this process, both by KIBS and by the client company.

The knowledge funnel also exists on the part of the client company (Fig. 5.3). Also, this site does not usually want to pass on too much knowledge to KIBS; this is especially true of knowledge of strategic importance. There are various reasons for

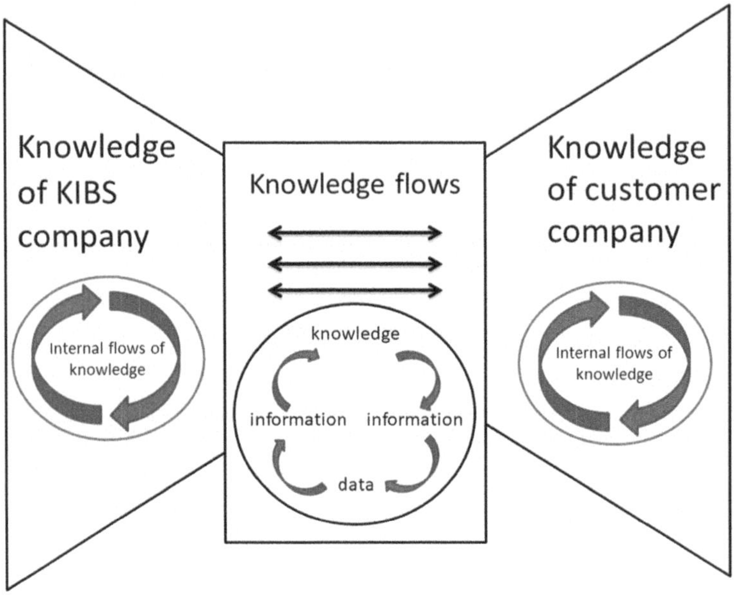

Fig. 5.4 Knowledge flows between KIBS and the client company. Source: Author's own illustration

limiting access to the knowledge of the customer company. First, the client company may fear knowledge leakage and loss of competitive advantage. By sharing knowledge about its operation with KIBS, the client company may be afraid that this knowledge will be used not only for the implementation of the KIBS service, but also for the needs of other companies—clients. Secondly, the client company is often not fully aware of what knowledge about its operation is important for the implementation of the KIBS service, and therefore may not see the need to provide it. Third, the client company may not be able to convey all the required knowledge or not be aware that this knowledge is available in the organization.

At the junction of the connection of these two funnels, knowledge flows between the KIBS company and the customer company (Fig. 5.4). In addition, internal knowledge flows at KIBS and the client company are usually required to capture and transfer the knowledge of each party. For example, to provide the KIBS service, it is necessary to obtain knowledge about the characteristics of customers and the market in which the customer company operates. To acquire this knowledge, the client company must acquire it from various departments of the company and prepare it in such a form that it can be passed on to the KIBS company that provides the service. The situation is similar in the KIBS company—the knowledge provided to the client should be collected and prepared in a form that will enable it to be transferred and absorbed by the client.

When knowledge is found in a given organization—both in the KIBS company and in the client company—it should flow between the people involved in the implementation of the KIBS service, as well as between these people and entire organizations. Thanks to this, better absorption and use of the obtained knowledge are possible.

On the one hand, KIBS companies integrate various areas of knowledge and competences in order to solve various problems of their clients. On the other hand, as already mentioned, knowledge flows between KIBS companies and their customers are not one-sided: KIBS companies also acquire knowledge from interactions with customers, thereby strengthening their own knowledge base. A three-phase model of knowledge processing as a result of this interaction has been proposed by Strambach (2001) and Muller and Zenker (2001) and includes the following activities:

- *knowledge acquisition*—KIBS companies acquire knowledge (hidden and overt) during the interaction process that takes place when the service is provided;
- *recombination of knowledge*—KIBS companies combine the knowledge acquired through collaboration with the existing knowledge, new knowledge is generated through this process of recombination/codification of knowledge, and the overall absorption capacity is increased and used in the next stage of knowledge processing;
- *dissemination of knowledge*—KIBS companies turn new knowledge into new products or service processes that open up new opportunities for them to interact and transfer knowledge to their customers.

The above activities constitute an iterative, continuous and reciprocal process in which interaction with the client expands the knowledge base of KIBS companies, which in turn results in new opportunities for interaction (Wong & He, 2002). Interaction with clients from the manufacturing and service sectors can also serve as a demand-pull factor for KIBS to constantly improve competences and keep up with the latest developments in its areas of knowledge (Strambach, 2001), e.g. in the field of information and communication technologies.

5.1.2 Knowledge Flows Between the Company Offering Knowledge-Intensive Business Services and the Client

KIBS may enter into various types of relationships with its clients, depending on the extent to which the company tailors the service to the client's needs and to what extent it must exchange knowledge with them when developing and delivering the service. Miles (2003) distinguished three types of relations between the client and the KIBS company:

- *sparring relationship*—in which the value of the service is negotiated between the supplier and the customer, communication is based on a balance in the level of knowledge and competence;

- *jobbing relationship*—where a lesser level of interaction is required and where the supplier is required to perform a specialized, technical task clearly defined by the customer and responding to a problem defined by the customer;
- *sales relationship*—in which the service is standardized and prepared before the transaction is concluded (Miles, 2003).

The type of customer relationship affects the knowledge flows between KIBS and its customers. A sparring relationship requires interaction and intensive knowledge exchange to the greatest extent. In the case of an employment relationship, standardization of the service offered reduces the need for interaction and knowledge exchange—the customer knows what he is buying, and KIBS does not have to tailor this service to his individual needs. The business relationship does not require intensive knowledge flows as it includes a standard service that does not require fitting and tailoring.

The relationship between KIBS and its client may have different phases: pre-relational phase, exploration phase, development phase and stable phase (Bagdoniene & Jakstaite, 2009). In the pre-relational phase, the client looks for a service provider who could help him solve the problem. At this stage, all the information that helps you evaluate individual suppliers and make the right choice is very valuable. In the exploration phase, the client contacts KIBS and starts a relationship. At this stage, both parties maintain a certain distance, because most often they do not have common experiences and have not yet exchanged knowledge about the essence of the problem and the possibilities of its solution. In the development phase, both parties increase the level of their knowledge as a result of an intensive exchange of knowledge and information, thus gaining a common understanding of problems and solutions. The parties get to know their expectations better and count on mutual benefits. In the last, stable phase, a long-term relationship is consolidated and deepened—both parties are already used to each other, know their ways of thinking and acting and have a common point of view regarding both the service and their own organizations (Bagdoniene & Jakstaite, 2009). It can be assumed that the flow of knowledge between the client and KIBS should be the most intensive in the development phase, because it is then that the creation of the service takes place, based on the mutual exchange of knowledge.

The relationship between KIBS and its customer from the knowledge exchange perspective is presented in Fig. 5.5, highlighting the interactions reflecting the different types of knowledge that are exchanged and combined in the service production process. The KIBS company must reach the customer and understand the knowledge about the processes and markets of its customers, as well as the difficulties and problems identified by the company—the customer, and then combine this with its own perception of technological or organizational solutions to the problem. This perception is based on professional knowledge and experience, including those gained in connection with the problems of other clients. For this reason, customers are ready to entrust KIBS with strategic information—it is necessary to create an appropriate solution to their specific problems, and also contributes to increasing KIBS's knowledge of various types of problems and

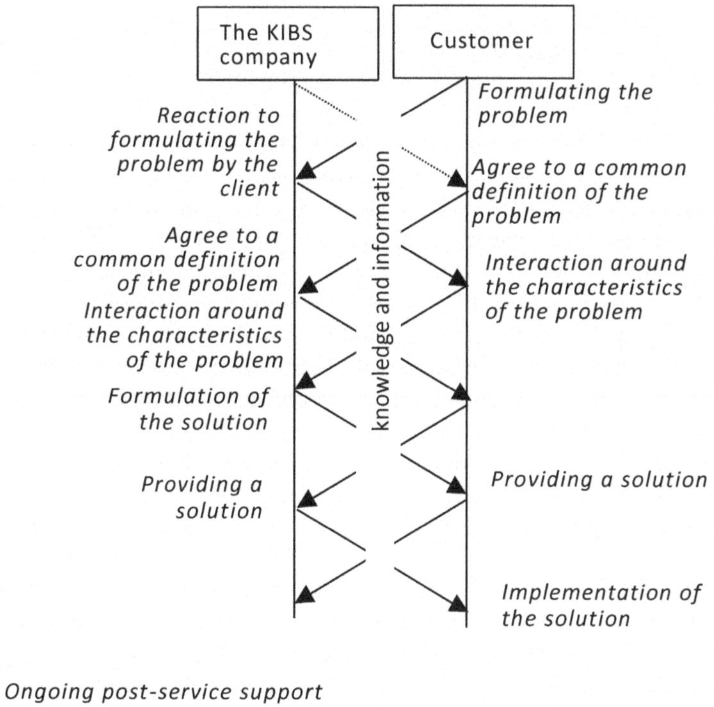

Fig. 5.5 Knowledge-sharing interactions between KIBS and its client. Source: Based on: Martinez-Fernandez and Miles (2006)

solutions. An important element of the KIBS service is the so-called post-service support, meaning the constant readiness of KIBS to help the client in using the provided solution (e.g. IT project) or constant customer care (e.g. in the case of companies offering accounting services). It should be assumed that this type of relationship, largely based on the exchange and flow of knowledge, will mainly concern the sparring relationship and, to some extent, the mercenary relationship.

Thus, it can be concluded that the previous publications on KIBS companies indicated a specific type of relationship between the KIBS company and its client and the role of acquiring and exchanging knowledge in this relationship. As research shows, when KIBS companies exchange knowledge with their clients, approx. 26% of this exchange concerns mainly classified knowledge (personalization strategy), approx. 19% almost exclusively or mainly codified knowledge (codification strategy) and approx. 55% combination of classified and codified knowledge (mixed strategy) (Landry et al., 2012). KIBS companies engage in knowledge exchange with clients for various strategic reasons that affect their day-to-day operations, as well as a portfolio of useful knowledge in future business situations. The types of knowledge exchanged between KIBS companies and their clients depend on the knowledge assets owned by KIBS, incl. the diversity of sources of knowledge, the creation of knowledge, knowledge possessed by employees, knowledge contained in management practices and technologies and the strength of the relationship between KIBS and customers. Based on a survey conducted among 1124 enterprises from the KIBS sector, the above-mentioned authors stated that:

1. The tendency of KIBS companies to rely more on the codification strategy and less on the personalization strategy increases with the number of different information sources, the number of knowledge workers, the diversity of knowledge management practices, the size of the company and the age of the company, and at the same time it decreases with the number of research investments—development, the variety of advanced technologies and the strength of ties.
2. KIBS companies operating in technology-based sectors, compared to KIBS companies operating in the sector of traditional professional services, more often use the codification strategy than personalization (Landry et al., 2012).

Summing up, it can be stated that the relations between KIBS companies and their clients are based on mutual flows of various types of knowledge, both explicit and hidden. In order to provide the KIBS service, the knowledge possessed not only by the service provider but also by the recipient of the service is necessary, and both parties must exchange this knowledge and fill their knowledge gaps. The role of knowledge flows will vary at different stages of the KIBS service provision and for different types of services, but it can be assumed that in most cases they will be simply essential. In order to improve the process of providing the KIBS service, the following questions can be asked at each stage of the service (Miles, 2012, p. 20):

1. What knowledge is required to successfully complete this service phase?
2. To what extent is this knowledge already possessed and to what extent must it be newly acquired?
3. To what extent does the company (KIBS or the client company) understand its own knowledge requirements and its partner's knowledge requirements in a service relationship?
4. Are there any incentives to acquire additional knowledge beyond that necessary for effective service delivery? Do these motivations go beyond the knowledge that may be helpful in future cases of creating a service relationship?
5. Does KIBS have skills that can be used to acquire knowledge and support a business partner in acquiring the necessary knowledge and to what extent?
6. What are the outcomes in terms of knowledge development and use in dealing with and using service providers?

Based on these questions, it can be concluded that if the results in terms of knowledge development and use are less than required for a given service, the service relationship will be a failure. To avoid this kind of situation, you should monitor the progress of the service relationship and develop catch-up methods. If the results in the form of development and use of knowledge turn out to be lower than expected, it may be a source of disappointment for one or both parties (KIBS and the client's companies), even if the service provided meets the formal requirements

(Miles, 2012, p. 20). That is why it is so important to study the issue of knowledge flows between KIBS companies and their customers and to identify how these flows can be improved. Failure to understand the concept of knowledge flows and the lack of awareness of the barriers they may encounter may hinder or even prevent KIBS companies from delivering their services.

KIBS companies use knowledge intensively—their activities are based on various types of exchange, knowledge flows both inside and outside the organization. Knowledge flows have a significant positive impact on innovation activities, in particular on the ability of companies to detect, adapt and use new knowledge and technology (Hipp, 2012, p. 149). In addition, KIBS companies have their own internal knowledge creation processes. Companies of this type must intensively acquire knowledge from various sources, as well as properly manage and use it. Taking this into account, it makes sense to examine the knowledge flows within KIBS companies and between KIBS companies and the external environment, as well as how knowledge is managed by these companies.

In relation to the above considerations, the following research questions were formulated:

- *What are the characteristics of knowledge flows in KIBS SMEs and between these companies and the environment?*
- *What knowledge flows within KIBS companies and between KIBS companies and their environment can be identified?*
- *What factors influence the knowledge flows within KIBS companies and between KIBS organizations and their surroundings?*
- *What tools are used for knowledge flows?*

5.1.3 Knowledge Management in Companies Offering Knowledge-Intensive Business Services

So far, little is known about how KIBS companies manage knowledge. Few publications deal with such issues as, for example, types of knowledge management strategies introduced by small- and medium-sized companies in this sector (Zieba et al., 2016) or the methods of knowledge absorption (Koch & Strotmann, 2008; Tseng et al., 2011) and the role of knowledge in creating innovation in the economy (Aslesen & Isaksen, 2007). Research was carried out on the impact of key success factors of KM on the innovativeness of KIBS companies, as well as selected aspects of knowledge management in such organizations.[4]

Based on the research conducted so far, it can be concluded that small- and medium-sized companies in the KIBS sector often choose the so-called an emergent strategy in the context of introducing knowledge management. This strategy can be defined as "informal, bottom-up initiatives in the field of knowledge management,

[4]For more information, see: Zieba and Zieba (2014) and Zieba (2014), pp. 1072–1079.

having its source in the daily practices and learning processes of the company's employees" (Zieba et al., 2016) and contrast with the *deliberate strategy*, which consists in purposeful and rational formulation of goals, plans and measures in the area of knowledge management by the company (Bolisani et al., 2016).

Knowledge, its absorption and management by KIBS companies are of great importance for the innovation of companies. As the results of the research conducted by Koch and Strotmann (2008) show, access to knowledge through networking and collaboration is crucial for innovative activity in the KIBS sector. Access to the knowledge of universities and research institutions is of particular importance for radical innovation, regardless of the degree of formalization of the collaboration. On the other hand, access to the knowledge of customers or suppliers has a significant impact on the likelihood of introducing innovation if the cooperation with these partners is based on formal cooperation through joint projects or formal cooperation agreements. The innovation performance of enterprises is significantly influenced by the contribution of knowledge—the higher the level of knowledge input, the better the innovation performance of enterprises (Tseng et al., 2011). Moreover, the innovativeness of KIBS sector companies is influenced by two important success factors when introducing KM: leadership and support from the management board, and incentive practices (Zieba & Zieba, 2014).

The author's research to date on knowledge management in Polish companies from the KIBS sector covers some fragmentary aspects of this issue, such as:

- KM practices implemented by these companies;
- KM practices introduced by small companies in this sector and success factors influencing the effectiveness of their implementation;
- KM practices introduced by companies from the SME sector KIBS and success factors influencing the effectiveness of their implementation.

The following conclusions can be drawn from these studies. As shown by the quantitative research conducted by the author, the vast majority of 112 surveyed companies introduced many practices related to knowledge management, with practices related to the implementation of information technologies being more popular than those related to human resource management. The explanation for this situation may be the fact that it is easier to introduce an IT solution in a company than to effectively manage human resources. Consequently, enterprises will be more inclined to implement "tangible", technical rather than "intangible" employee management practices. The least frequently introduced practice was the development and implementation of a knowledge management strategy—only over a third of companies declared to implement this type of strategy (Zięba, 2015b). Regarding the success factors in introducing KM practices, the research conducted by the author indicated the following:

- *factors related to human resource management*—dialogue with the employee, listening to his ideas and reasons, knowledge exchange between employees,

management board leadership, communication between the management board and employees, motivating and encouraging employees, employee openness;
- *factors related to technical aspects*—security and ease of access to information;
- *business factors*—avoidance of routine and the ability to forget past experiences, profitability and business case.

Despite the availability of the results of the research presented above, the issue of knowledge management in KIBS companies is still not sufficiently researched. For example, there is still a lack of analyses showing the relationship between the HR practices implemented by such organizations and their impact on the results achieved by them. The only available study on the impact of knowledge management on the performance of this type of organization was conducted by Lara et al. (2012) and included Spanish companies with over 100 employees. This issue has not been studied among smaller KIBS companies and companies located in other countries (including Poland). Therefore, it is legitimate to try to answer the following research questions:

- *How do companies from the KIBS SME sector manage knowledge?*
- *What kind of solutions/practices in the area of KM are introduced and used by these companies?*
- *What are the reasons for introducing or not introducing particular practices?*

5.2 Methodological Foundations of the Research and Description of the Research Method

Social reality can be addressed in various ways, and researchers may adopt various ontological and epistemological attitudes, which are reflected in their choice of methodology and research approach. These elements are very important to research as they shape the approach to theory and methods used; they are also deeply rooted in the researcher's beliefs about the world (Gupta & Awasthy, 2015, p. 5). In general, the researcher's approach to reality can be described as a paradigm, understood as a set of basic concepts and theories that form the basis of a given science (Kuhn, 2001).

In the case of the studied phenomenon, the application of the interpretative paradigm seems justified for several reasons. First, the objectives of the study were to develop and verify the theory of knowledge and its flows in KIBS SMEs, and not to test hypotheses and create statistical generalizations based on available constructs. According to Sulkowski (2009, p. 128), "interpretation processes give meaning to the environment and form it in cognizable and formable beings. The interpretative approach stresses the feedback between the discovery and creation of the world by man in the process of cognition" (Sulkowski, 2009). In this case, the results of the study will be an intersubjective product of its participants (KIBS SMEs and their broadly understood environment). The purpose of building a theory in the interpretive paradigm is to generate descriptions and explanations of events so that the system of interpretation and meaning, as well as the processes of structuring and organizing (Gioia & Pitre, 1990). Participants in this reality perceive it as obvious, but the researcher's task is to show how it was created (interpretation). Secondly, the aim of interpretative analysis is to reach things important for social actors, but often not subject to discussion and taken for granted (Jemielniak, 2008, p. 56). In the case of research on phenomena related to knowledge in the context of KIBS SMEs, many aspects can be perceived as obvious, and at the same time it is important to reach deeper meanings and explain the reasons for this state of affairs. Thirdly, the interpretative paradigm is underestimated in management sciences, which makes it less popular compared to the functional paradigm, and at the same time it can provide a lot of valuable information on the functioning of the organization (Jemielniak, 2008, p. 56). Fourth, in interpretative research, the researcher is very close to the analysed issue and analyses the results of the research in a subjective way (Gupta & Awasthy, 2015, p. 12). The author of this work is involved in researching issues related to KIBS companies due to her long-term interest in these entities. Interpretation of phenomena is not objective, but allows for a thorough understanding of the specifics of KIBS companies. The author's earlier research allowed to identify the research gap and to understand the specificity of the functioning of KIBS companies.

In order to maintain the methodological rigour, the study was conducted in accordance with the template for conducting research under the interpretative paradigm proposed by Hallebone and Priest (2009). First, the research topic, its purpose and potential application are outlined. The second phase focused on potentially useful knowledge and the main research question that helped guide the main research strategy. The next two steps are: data generation and simultaneous analysis as well as concept generation and validation. These stages intertwined with each other. In the final phase, the research results and their potential application were developed. As part of generating data, in-depth interviews were conducted with respondents in order to examine concepts from literature and learn about concepts that result from the reported experiences of participants and the meanings they derive from them (Hallebone & Priest, 2009, pp. 55–59). A detailed description of the approach used is provided in the next subchapter.

If an interpretative paradigm is adopted, qualitative methods are usually the natural choice for the research methods used (Hallebone & Priest, 2009, p. 56). Qualitative research is a form of social science that focuses on understanding the world, interpreting people's experiences and finding meaning in them. This kind of research is about the subjective world we live in and is based on seeing this world in a holistic way and with many constructed, existing realities. Behaviour is conditioned by the situation and the interpretation of the context (Gupta & Awasthy, 2015, p. 13). Moreover, qualitative research offers the possibility of obtaining the most faithful (uninterrupted by the person conducting the research) description of a segment of social reality. Their goals include understanding, enlightenment and—if possible—extrapolation of results to similar situations. Qualitative research enables the creation of comprehensive knowledge, defining phenomena in their natural context. Among the qualitative research strategies proposed by Creswell (2013), i.e. ethnography, grounded theory, case studies, phenomenological research and narrative research (Creswell, 2013, pp. 38–39), the case study strategy was selected as the most appropriate. First, case studies are especially useful in the early stages of a scientific problem diagnosis. The issue of knowledge flows is insufficiently defined and researched in the literature, and therefore requires exploratory research. Secondly, the case study method is appropriate when the question concerns a contemporary (non-historical) event or phenomenon that is beyond the researcher's control or influence is negligible, and contains "how" or "why" (Yin, 2009, p. 13). The formulated research questions are of this kind. Third, this method is the more appropriate, the more the research question requires a comprehensive and in-depth description of a social phenomenon. The case study method allows for theoretical research, theory testing as well as application applications. Fourth, this method allows both the phenomena of interest and context to be explored, creating a large number of potentially relevant elements (Yin, 2009). Additionally, in the field of knowledge management qualitative methods are often applied.

Much preparation is required to identify the most important questions for a given research topic and to be precise in the formulation of these questions. One way is to review the literature on the subject. Experienced researchers analyse previous research to develop more precise and insightful questions on a given topic (Yin, 2009, p. 14). This method was used in this monograph—a review of the available literature and research conducted by the author allowed for the formulation of detailed research questions. It needs to be kept in mind that the case study analysis does not offer the possibility of generalizing the results of the obtained study to the entire population, but only allows for the formation of theoretical propositions (Yin, 2009, p. 14). That is how this method will be applied.

5.3 Study Description

In the previous studies, the author has addressed issues related to companies in the knowledge-intensive business services sector, both at the macroeconomic level (Zieba, 2013; Zięba, 2015a), and at the level of individual companies (Zieba & Zieba, 2014; Zieba & Kończyński, 2017; Zięba, 2014, 2015a). While examining the critical success factors in knowledge management by small- and medium-sized companies offering knowledge-intensive business services,[5] a research gap was identified regarding the flow of knowledge in companies in this sector.

The main objectives of the case study research included *explaining the importance of knowledge for small- and medium-sized companies offering knowledge-intensive business services (KIBS SMEs) and its impact on the competitive advantage of these companies, as well as a comprehensive discussion of the phenomenon*

[5]The study was conducted as part of a research grant financed by the National Science Center, grant no. 2011/01/D/HS4/04111, entitled Knowledge management in small and medium-sized enterprises offering knowledge-intensive business services.

of knowledge flows taking place both inside KIBS SMEs companies and between KIBS SMEs companies and their broadly understood surroundings.

In order to achieve the above goals, the following research questions were formulated:

1. What knowledge is important for the functioning of KIBS SMEs?
2. What knowledge do companies in this sector perceive as the most important for their activities?
3. How does knowledge affect the competitive advantage of these companies?
4. What are the characteristics of knowledge flows in KIBS SMEs and between these companies and the environment?
5. What knowledge flows inside KIBS companies and between KIBS companies and their environment can be identified?
6. What factors influence the knowledge flows within KIBS companies and between KIBS organizations and their surroundings?
7. What tools are used for knowledge flows?
8. How do companies from the KIBS SME sector manage knowledge?
9. What kind of solutions/practices in the area of KM are introducing and applying?
10. What are the reasons for introducing or not introducing particular practices?

The procedure for conducting the study was developed in accordance with the guidelines proposed by Yin (2009) and Czakon (2013). In the first stage, an in-depth review of the literature on KIBS companies and knowledge management in organizations was carried out. On the basis of this review, a cognitive gap was identified resulting from the available research and observation of reality. This gap was the basis for the formulation of research questions. Then, case studies were selected for the study and data collection tools were developed. After conducting research in KIBS SMEs companies and collecting data, individual case studies were analysed, followed by a cross-sectional analysis.[6] Although the procedure for conducting the qualitative study was linear and the steps followed one another, the steps followed many times—this type of iteration is one of the requirements of a case study research. The procedure for conducting the qualitative test is shown in Fig. 5.6.

The companies for the study were selected on the basis of a database prepared on request by Infobrokering and covering companies from the PKD 2007 subsectors, which were considered to belong to the KIBS sector (Chap. 2), i.e. divisions 62, 63, 69, 70, 71, 72, 73, 74 and 78 (Schnabl & Zenker, 2013). Reaching companies offering various types of services allowed for the acquisition of more diverse empirical material, as well as for comparing and contrasting the functioning of

[6]This chapter presents only the results of the cross-sectional analysis of cases, for two reasons: (1) describing 14 individual cases could tire the reader; (2) the companies participating in the study did not agree to make public much of their data, which would allow the identification of the surveyed entities. Therefore, this analysis has been performed, but it has not been included in this study.

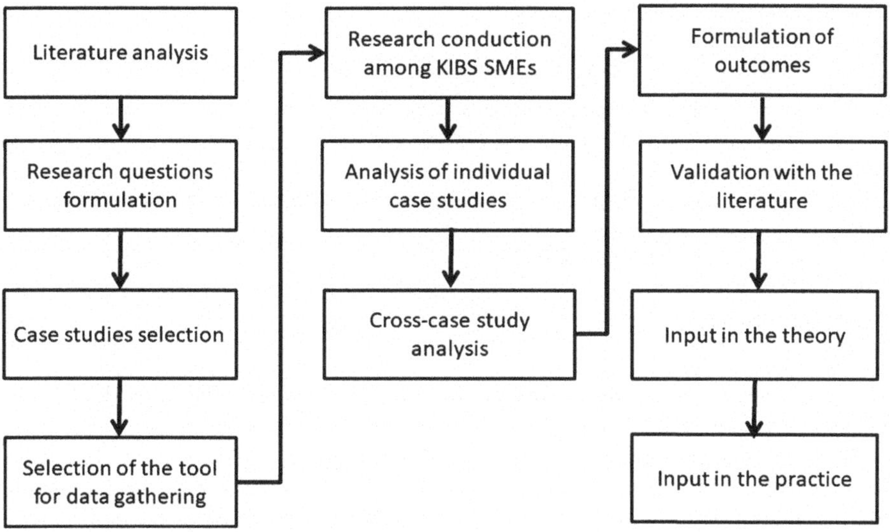

Fig. 5.6 Procedure for conducting qualitative research. Source: Author's own illustration, based on: Yin (2009), p. 57 and Czakon (2013), p.102

KIBS companies. The vast majority of organizations included in the database were micro and small companies (92.5%), therefore they were mainly selected for the study. The small share of medium-sized entities in the population (16 out of 246 companies of all companies of this type in the acquired database, which constituted 6.5%) is reflected in the study: one average organization in the population of 14 companies constitutes 7% of this number.

Firms that refused to participate in the study cited various reasons, such as:

- lack of time;
- lack of interest in the research subject;
- reluctance to participate in any research;
- the company's privacy policy, according to which they do not share any information about themselves (as already mentioned, this reason was often mentioned in the case of medium-sized companies);
- lack of a person who could answer the questions related to the subject of the study.

The time scope of the study covered the period from September to December 2017. The surveyed entities were small- and medium-sized companies offering knowledge-intensive business services. The subject of the study was knowledge and knowledge flows in companies of this type, as well as between these companies and their broadly understood environment. The territorial scope of the study included the Tri-City region of Poland (Gdańsk, Sopot and Gdynia) due to the

Table 5.2 Individual scopes of the study

Type of range	Description of the scope
Time range	From September to December 2017
Subject scope	KIBS small- and medium-sized companies
Objective scope	Knowledge and flow of knowledge in KIBS SMEs and between these companies and the wider environment
Territorial scope	Tri-City (Region of Poland)

Source: Own elaboration

convenient access, allowing to shorten the time of the study. The individual study ranges are presented in Table 5.2.

The selection of companies was layered, non-random, purposeful and based on the availability of the surveyed entities. The layers concerned the size of companies (micro, small, medium) and the type of services they offer (type of business). This type of approach has also been used in other studies on KIBS companies (Teirlinck & Spithoven, 2013) and companies from the SME sector (Chong et al., 2011; Wang & Yang, 2016). It should also be noted that convenience sampling is often used in management research (Quaddus & Xu, 2005). A total of 14 companies from the KIBS SME sector were selected for the study, offering various types of services (including consultancy, legal, accounting, advertising, surveying, information technology, research and development) and employing a diverse number of employees (Table 5.3).

The structure of the surveyed companies in relation to their size was as follows: 5 micro companies, 8 small companies and 1 medium company. It is worth noting that one of the companies (Company H) employed 40 employees, so its size was close to a medium-sized company. The number of surveyed companies was limited by the issues of access to the surveyed entities, available resources and the time devoted to research, although a similar number of cases was also studied by other authors dealing with the area of knowledge management (e.g. Nunes et al., 2006; Bishop et al., 2008).

The study of case studies is usually associated with a variety of phenomena and a significant complexity of the context of the studied reality, which the researcher has to deal with. Therefore, it is necessary to collect data from various sources and triangulate the results. In practice, triangulation requires the use of several different procedures, which allows for greater certainty in the process of searching for answers to research questions. In order to authenticate the conducted research, the method of data triangulation was used, which enables obtaining a broader context of the studied phenomenon, ensuring higher quality of research and eliminating measurement errors related to relying on one data source. The analysis of data collected from various sources allowed for the verification of the information provided and enriched the overall picture of the phenomena studied.

Table 5.3 Characteristics of the surveyed companies

Firm	Offered services	No. of employees	Date of establishment	Person interviewed
Firm A	Tax advisory services and accounting and accounting services	13	1990	Owner
Firm B	Advertising services	9	2005	Owner
Firm C	Services for the production of dedicated software	15	2015	Owner
Firm D	Software development and implementation services	12	1992	Owner
Firm E	Advertising services	7	2006	Owner
Firm F	Accounting and accounting and management services	7	1999	Manager
Firm G	Technical design services	20	1992	Owner
Firm H	Technical design services	40	1992	President of the board
Firm I	Accounting services	5	2003	Owner
Firm J	Accounting and bookkeeping services	20	2004	Manager
Firm K	Geodetic services (the legal aspect)	6	1995	Owner
Firm L	Scientific and research services	70	1954	Manager
Firm M	Advertising services	20	2007	Owner
Firm N	Geological, geotechnical and geohydrological services	20	1988	Technical director

Source: Own elaboration

Both the data collection procedures and the procedures for their documentation were selected in line with the recommendations of Creswell (2013, pp. 194–199). The following data collection procedures were used for the study:

- partially structured face-to-face interviews;
- observations made during the interview, before its commencement and after its conduct;
- websites of the surveyed companies and their accounts in social networks (FB, LinkedIn);
- other materials (e.g. a drawing made by the respondent, available press articles on some companies).

The following data documentation procedures were used:

- interview protocol;
- written notes from the interview with the observations made—notes on, inter alia, company headquarters, interior, furniture, atmosphere, clothing, activities that could be observed during the interview (e.g. conversation with a client),

impressions from the meeting, items displayed (e.g. publications in the field of specialist literature), etc.;
- keeping a diary for the duration of the study;
- accurate transcriptions of interviews.

The owners or managers of the surveyed companies were invited to the study, because in the case of small- and medium-sized companies they are perceived as key informants and a similar approach was used in research related to knowledge management in such companies (Desouza & Awazu, 2006; Palacios-Marqués et al., 2013; Wong & Aspinwall, 2005). Moreover, in the case of small- and medium-sized enterprises, owners and managers are often well versed in the activities undertaken by employees and the way the organization operates, as they themselves take part in the daily tasks. In the case of 9 companies, the owner (or owner) participated in the research, in 2 cases it was a manager and in the remaining 3 cases it was the president of the board, technical director and specialist. Before each interview, the respondents were presented with the purpose of the study and information on the confidentiality of the collected materials. Before the start of the study, each of the participants gave their consent in writing to conduct it and to record the interview (consent could be withdrawn at any time during the interview, but none of the respondents took advantage of this option). The identities of interviewees were coded to preserve the anonymity of the surveyed companies. Then the interviews were carefully written down, keeping all the details, such as sighs, repetitions, exclamation marks, etc., and also verified with other obtained materials, such as research notes or information taken from the websites of the surveyed companies. All these elements allowed for the validation of the obtained results (Suter, 2011). The interviews lasted from 20 to 60 min—some respondents answered the questions very briefly and did not develop their statements, while others broadened individual responses.

The obtained data (both from interviews and other sources) was coded and analysed with the support of qualitative analysis software (NVIVO 11 Professional). The analysis was carried out on the basis of the stages proposed by Creswell (2013, pp. 201–205).

In the first stage, the data for analysis were organized and prepared, namely: the interviews were transcribed, the paper materials were scanned, the content of websites and social profiles was saved in PDF version, handwritten notes collected during the research were entered into the computer and the data was organized by dividing them into different categories. In a second step, all data was reviewed to gain an overview and to reveal the common meaning of the information obtained. The third step was to start the detailed analysis with the encoding process. The study used the developed theoretical framework presented in the previous chapters, therefore, when creating the analytical codes, it referred to research questions and the adopted theoretical perspective. The coding was therefore partially closed. At the same time, the data that did not match the created codes formed the basis for creating open codes. After the collected information was sorted according to categories, it was analysed within individual categories and the relations between them. For this purpose, the technique of cross-case analysis was chosen, organizing the data from individual cases according to the same code categories. In the fourth stage, the coding process was used to describe the analysed phenomena, and in the next stage, the method of presenting the description and thematic scopes was determined. In order to summarize the obtained and analysed results, it was decided to use tables. In the last stage, an interpretation was made, that is, the meaning of the obtained results was explained, trying to answer the question: what does it result from? The obtained results were also confronted with the literature on the subject and the theories proposed in the previous subsections.

When presenting the research results, it was decided to use quotes from interviews to a large extent, because they constitute important empirical material and show the way of thinking of representatives of KIBS companies. This way of presenting the results of the research has some imperfections (e.g. the use of colloquial language), but it allows for a better understanding of the reality of the functioning of KIBS companies. The results of the study, along with their description and interpretation, are included in the following chapters. For each of the discussed issues, quotes illustrating the individual approaches of the respondents along with coding are presented.

5.4 Summary

Considering the current lack of research on the concept of knowledge flows in KIBS SMEs and the high complexity of this issue, ensuring methodological rigour is a particularly important element. Therefore, the purpose of this chapter was to present not only theoretical aspects and individual stages of the research, but also their logical justification, while being aware that the separation of knowledge flows in the organization and their analysis is a complex task. Potential problems undoubtedly include: subjective perception of the issue of flows by business owners, difficulties in defining knowledge affecting the definition of its flow or difficulties in distinguishing between flows of knowledge and flows of information and data.

In the first part, this chapter presented the philosophical foundations of the research, which then affected the method used. The second part presented the description of the study: its main assumptions, the purpose of the study together with research questions, the research method and the individual steps of the research procedure. Detailed criteria for selecting cases for the study were also discussed, followed by the methods of collecting and analyzing data.

References

Amara, N., Landry, R., & Traoré, N. (2008). Managing the protection of innovations in knowledge-intensive business services. *Research Policy, 37*(9), 1530–1547. https://doi.org/10.1016/j.respol.2008.07.001

Aslesen, H. W., & Isaksen, A. (2007). New perspectives on knowledge-intensive services and innovation. *Geografiska Annaler, 89B*(S1), 45–58.

Bagdoniene, L., & Jakstaite, R. (2009). Trust as basis for development of relationships between professional service providers and their clientes. *Economics and Management, 14*, 360–366.

Bishop, J., Bouchlaghem, D., Glass, J., & Matsumoto, I. (2008). Ensuring the effectiveness of a knowledge management initiative. *Journal of Knowledge Management, 12*(4), 16–29. https://doi.org/10.1108/13673270810884228

Bolisani, E., Scarso, E., & Zieba, M. (2016). How to deal with knowledge in small companies? Defining emergent Km approach. *International Journal of Learning and Intellectual Capital, 13* (2–3), 104–118. https://doi.org/10.1504/IJLIC.2016.075701

Bukowitz, W. R., & Williams, R. L. (2000). *The knowledge management fieldbook*. Financial Times/Prentice Hall.

Chong, C. W., Chong, S. C., & Gan, G. C. (2011). Inter-organizational knowledge transfer needs among small and medium enterprises. *Library Review, 60*(1), 37–52. https://doi.org/10.1108/00242531111100568

Creswell, J. W. (2013). *Projektowanie badań naukowych: metody jakościowe, ilościowe i mieszane*. Wydawnictwo Uniwersytetu Jagiellońskiego.

Czakon, W. (Ed.). (2013). *Podstawy metodologii badań w naukach o zarządzaniu*. Oficyna Wolters Kluwer. https://doi.org/10.1007/s13398-014-0173-7.2

Dalkir, K. (2011). *Knowledge management in theory and practice*. MIT Press.

Davenport, T. H., & Prusak, L. (1998). *Working knowledge: How organizations manage what they know*. Harvard Business Press.

Desmarchelier, B., Djellal, F., & Gallouj, F. (2013). Knowledge intensive business services and long term growth. *Structural Change and Economic Dynamics, 25*(1), 188–205. https://doi.org/10.1016/j.strueco.2012.07.003

Desouza, K. C., & Awazu, Y. (2006). Knowledge management at SMEs: Five peculiarities. *Journal of Knowledge Management, 10*(1), 32–43.

Gallouj, F. (2002). Knowledge-intensive business services: Processing knowledge and producing innovation. In J. Gadrey & F. Gallouj (Eds.), *Productivity, innovation and knowledge in services* (pp. 256–284). Edward Elgar.

Gibbert, M., Leibold, M., & Probst, G. (2002). Five styles of customer knowledge management, and how smart companies use them to create value. *European Management Journal, 20*(5), 459–469. https://doi.org/10.1016/S0263-2373(02)00101-9

Gioia, D. A., & Pitre, E. (1990). Multiparadigm perspectives on theory building. *Academy of Management Review, 15*(4), 584–602. https://doi.org/10.5465/AMR.1990.4310758

Gupta, R. K., & Awasthy, R. (2015). In R. K. Gupta & R. Awasthy (Eds.), *Qualitative research in management. Methods and experiences*. Sage Publications.

Hallebone, E., & Priest, J. (2009). *Business & management research*. Palgrave Macmillan.

Hamel, G., & Prahalad, C. K. (1999). *Przewaga konkurencyjna jutra: Strategie przejmowania kontroli nad branżą i tworzenia rynków przyszłości*. Business Press.

Hipp, C. (2012). Information flows and knowledge creation in knowledge-intensive. In J. S. Metcalfe & I. Miles (Eds.), *Innovation systems in the service economy: Measurement and case study analysis* (Vol. 18, pp. 149–168). Springer Science & Business Media.

Jashapara, A. (2011). *Knowledge management. An integrated approach*. Pearson Education.

Jemielniak, D. (2008). *Praca oparta na wiedzy. Praca w przedsiębiorstwach wiedzy na przykładzie organizacji high-tech*. Wydawnictwa Akademickie i Profesjonalne.

Koch, A., & Strotmann, H. (2008). Absorptive capacity and innovation in the knowledge intensive business service sector. *Economics of Innovation and New Technology, 17*(6), 511–531. https://doi.org/10.1080/10438590701222987

Kohtamäki, M., & Partanen, J. (2016). Co-creating value from knowledge-intensive business services in manufacturing firms: The moderating role of relationship learning in supplier–customer interactions. *Journal of Business Research, 69*(7), 2498–2506. https://doi.org/10.1016/j.jbusres.2016.02.019

Kuhn, T. S. (2001). *Struktura rewolucji naukowych*. Fundacja Aletheia.

Landry, R., Amara, N., & Doloreux, D. (2012). Knowledge-exchange strategies between KIBS firms and their clients. *The Service Industries Journal, 32*(March 2015), 291–320. https://doi.org/10.1080/02642069.2010.529131

Lara, F., Palacios-Marques, D., & Devece, C. A. (2012). How to improve organisational results through knowledge management in knowledge-intensive business services. *Service Industries Journal, 32*(11), 1853–1863. https://doi.org/10.1080/02642069.2011.574283

Leonard-Barton, D. (1995). *Wellsprings of knowledge: Building and sustaining the sources of innovation*. Harvard Business School Press.

Martinez-Fernandez, M. C., & Miles, I. (2006). Inside the software firm: Co-production of knowledge and KISA in the innovation process. *International Journal of Services, Technology and Management, 7*(2), 115–125. http://www.scopus.com/inward/record.url?eid=2-s2.0-33745800228&partnerID=tZOtx3y1

Miles, I. (2012). KIBS and knowledge dynamics in client-supplier interaction. In E. Di Maria, R. Grandinetti, & B. Di Bernardo (Eds.), *Exploring knowledge-intensive business services: Knowledge management strategies* (pp. 13–34). Palgrave.

Miles, I. (2003). Knowledge-intensive services' suppliers and clients. In *Studies and reports*. Ministry of Trade and Industry.

Muller, E., & Zenker, A. (2001). Business services as actors of knowledge transformation: The role of KIBS in regional and national innovation systems. *Research Policy, 30*(9), 1501–1516. https://doi.org/10.1016/S0048-7333(01)00164-0

Murray, P., & Myers, A. (1997). The facts about knowledge. *Information Strategy, 2*(7), 29–33.

Nonaka, I. (1994). A dynamic theory of organizational knowledge creation. *Organization Science, 5*(1), 14–37. https://doi.org/10.1287/orsc.5.1.14

Nonaka, I., Toyama, R., & Konno, T. (2000). SECI, Ba and leadership: A unified model of dynamic knowledge creation. *Long Range Planning, 33*.

Nunes, B. M., Annansingh, F., Eaglestone, B., & Wakefield, R. (2006). Knowledge management issues in knowledge-intensive SMEs. *Journal of Documentation, 62*(1), 101–119. https://doi.org/10.1108/00220410610642075

Päällysaho, S., & Kuusisto, J. (2011). Informal ways to protect intellectual property (IP) in KIBS businesses. *Innovation: Management, Policy & Practice, 13*(March 2015), 62–76. https://doi.org/10.5172/impp.2011.13.1.62

Palacios-Marqués, D., Peris-Ortiz, M., & Merigó, J. M. (2013). The effect of knowledge transfer on firm performance: An empirical study in knowledge-intensive industries. *Management Decision, 51*(5), 973–985. https://doi.org/10.1108/MD-08-2012-0562

Paquette, S. (2011). Customer knowledge management. In D. Schwartz & D. Te'eni (Eds.), *Encyclopedia of knowledge management* (2nd ed., pp. 175–184). IGI Global.

Pardos, E., Gómez-Loscos, A., & Rubiera-Morollón, F. (2007). 'Do versus Buy' decisions in the demand for knowledge intensive business services. *The Service Industries Journal, 27*(3), 233–249. https://doi.org/10.1080/02642060701206991

Perrott, B. E. (2007). A strategic risk approach to knowledge management. *Business Horizons, 50* (6), 523–533. https://doi.org/10.1016/j.bushor.2007.08.002

Probst, G., Romhardt, K., & Raub, S. (2000). *Managing knowledge: Building blocks for success*. J. Wiley.

Quaddus, M., & Xu, J. (2005). Adoption and diffusion of knowledge management systems: Field studies of factors and variables. *Knowledge-Based Systems, 18*(2–3), 107–115. https://doi.org/10.1016/j.knosys.2004.11.001

Schnabl, E., & Zenker, A. (2013). Statistical classification of knowledge-intensive business services (KIBS) with NACE Rev. 2. *EvoREG Research Note, 25*(June), 1–11. papers3://publication/uuid/7AB446C4-BC91-44AF-8EB2-A5BC5950EB04.

Strambach, S. (2001). Innovation processes and the role of knowledge-intensive business services (KIBS). In K. Koschatzky, M. Kulicke, & A. Zenker (Eds.), *Innovation networks: Concepts and challenges in the European perspective* (pp. 53–68). Physica-Verlag HD. https://doi.org/10.1007/978-3-642-57610-2_4

Sulkowski, L. (2009). Interpretative approach in management sciences. *Argumenta Oeconomica, 23*(2), 127–149.

Suter, W. N. (2011). *Introduction to educational research: A critical thinking approach*. SAGE Publications.

Teirlinck, P., & Spithoven, A. (2013). Formal R&D management and strategic decision making in small firms in business services. *R&D Management, 43*(1), 37–51. https://doi.org/10.1111/j.1467-9310.2012.00701.x

Tseng, C.-Y., Pai, D. C., & Hung, C.-H. (2011). Knowledge absorptive capacity and innovation performance in KIBS. *Journal of Knowledge Management, 15*(6), 971–983. https://doi.org/10.1108/13673271111179316

Tuomi, I. (1999). Data is more than knowledge: Implications of the reversed knowledge hierarchy for knowledge management and organizational memory. *Journal of Management Information Systems, 16*(3), 103–117. https://doi.org/10.1109/HICSS.1999.772795

Tuominen, T., & Toivonen, M. (2011). Studying innovation and change activities in KIBS through the lens of innovative behaviour. *International Journal of Innovation Management, 15*(2), 393–422. https://doi.org/10.1142/S1363919611003209

Wang, M.-H., & Yang, T.-Y. (2016). Investigating the success of knowledge management: An empirical study of small- and medium-sized enterprises. *Asia Pacific Management Review, 21*(2), 79–91. https://doi.org/10.1016/j.apmrv.2015.12.003

Wong, K. Y., & Aspinwall, E. (2005). An empirical study of the important factors for knowledge-management adoption in the SME sector. *Journal of Knowledge Management, 9*(3), 64–82. https://doi.org/10.1108/13673270510602773

Wong, P., & He, Z. (2002). Discussion Paper No. 2002 / 69 the impacts of knowledge interaction with manufacturing clients on KIBS firms innovation behaviour. *Knowledge Creation Diffusion Utilization*, 10–11.

Yin, R. K. (2009). Case study research: Design and methods. In L. Bickman & D. J. Rog (Eds.), Essential guide to qualitative methods in organizational research (Vol. 5, 5). Sage Publications. https://doi.org/10.1097/FCH.0b013e31822dda9e

Zięba, M. (2014). Zarządzanie wiedzą w małych przedsiębiorstwach oferujących wiedzochłonne usługi biznesowe. *Prace Naukowe Wyższej Szkoły Bankowej w Gdańsku, 37*, 163–172.

Zięba, M. (2015a). Jak firmy z sektora wiedzochłonnych usług biznesowych zarządzają wiedzą? Prezentacja wyników badań. *Studia i Prace Wydziału Nauk Ekonomicznych i Zarządzania, 39*(4), 461–473.

Zięba, M. (2015b). Sektor usług wiedzochłonnych oraz jego dynamika i struktura zatrudnienia w krajach Unii Europejskiej. Ekonomia. *Rynek, Gospodarka, Społeczeństwo, 40*, 133–153.

Zięba, M. (2017). Knowledge safety – Insights from the SME sector. *Journal of Management and Business Administration. Central Europe, 25*(3), 78–96. https://doi.org/10.7206/jmba.ce.2450-7814.203

Zieba, M., & Zieba, K. (2014). Knowledge management critical success factors and the innovativeness of KIBS companies. *Inzinerine Ekonomika-Engineering Economics, 25*(4), 458–465. https://doi.org/10.5755/j01.ee.25.4.6494

Zieba, M. (2013). Knowledge-intensive business services (KIBS) and their role in the knowledge-based economy. *Proceedings of the European Conference on Knowledge Management, ECKM, 2*, 785–792.

Zieba, M. (2014). Critical success factors for knowledge management in SMEs in the KIBS sector. *European Conference on Knowledge Management, 2*, 1072–1079.

Zieba, M., Bolisani, E., Paiola, M., & Scarso, E. (2017). Searching for innovation knowledge: Insight into KIBS companies. *Knowledge Management Research & Practice, 15*(2). https://doi.org/10.1057/s41275-017-0050-1

Zieba, M., Bolisani, E., & Scarso, E. (2016). Emergent approach to knowledge management by small companies: Multiple case-study research. *Journal of Knowledge Management, 20*(2), 292–307. https://doi.org/10.1108/JKM-07-2015-0271

Zieba, M., & Kończyński, P. (2017). Client co-production in knowledge-intensive business services (KIBS): Case study analysis. *Proceedings of the European Conference on Knowledge Management, ECKM, 2*, 1075–1083.

Knowledge and Its Flows in the Light of Empirical Research 6

6.1 Characteristics of the Surveyed Companies

First, the surveyed companies were asked about several aspects related to the type of services they offer to their clients. In addition to describing the services offered, questions were also asked about their routine. In general, KIBS are perceived in the literature as non-routine (Corrocher et al., 2009; Muller & Doloreux, 2009), but it does not always have to be this way. Certain services may consist of standard components that are tailored to the customer's needs. Routine or standardized services may determine the intensity of knowledge flows. For example, when KIBS offer standardized services, the knowledge flows between KIBS and the client will not be very intense, while when delivering a service that is tailored to the client's needs, intensive knowledge flows will be a must. Without the flow of knowledge, it will not be possible to refine the subject of the service or adjust the solution to real needs. In other words, it can be expected that the more non-standard/unconventional a service is to be, the more intense the knowledge flows related to its implementation must be. First, the types of services offered by the surveyed companies are described and then their routine.

6.1.1 Services Offered

The surveyed companies belonged to different KIBS subsectors and offered a diverse spectrum of services. Each company was asked to briefly describe what it is doing to better illustrate the type of business. The mere use of NACE codes in research is insufficient in the sense that organizations often change or expand their activities, and the NACE codes remain the same as when the company was founded.

Detailed information on the services offered is presented in Table 6.1.

As shown in Table 6.1, the survey covered companies offering:

M. Zieba, *Understanding Knowledge-Intensive Business Services*, Knowledge Management and Organizational Learning 10,
https://doi.org/10.1007/978-3-030-75618-5_6

Table 6.1 Description of services offered by the surveyed KIBS companies

Firm	Offered services	Detailed description of the services
Firm A	Tax advisory services as well as bookkeeping and accounting services	Providing advice and explanations in the field of tax and customs obligations, preparing opinions on tax and customs matters, social insurance institution contributions, fiscal and administrative fees, tax optimization, tax audit, representation before public administration authorities and administrative courts, accounting and accounting services, settlements with social insurance institution
Firm B	Advertising services	Broadly understood advertising service, e.g. logo designs, printing services, advertising services, as well as production and assembly
Firm C	Dedicated software development services	IT services—custom software development
Firm D	Software development and implementation services	Creating software for sales support and accompanying services, such as: Implementation of this software, sale of equipment that is needed to implement comprehensive systems, service of supplied equipment and consulting
Firm E	Advertising services	Small website designs, extensive portals, web applications, e-commerce services, customer marketing support, website building, online advertising, hosting, domains, visual identification
Firm F	Accounting and accounting and management services	Comprehensive accounting and tax services for small- and medium-sized business entities
Firm G	Technical design services	Ship design
Firm H	Technical design services	Engineering activities and related technical consultancy
Firm I	Accounting services	Comprehensive tax and accounting services for small- and medium-sized companies, consulting in the field of business activities, representation in state institutions, payroll services, supervision over personnel documentation, preparation of reports, loan applications, financial analyses, preparation of various types of reports and documentation
Firm J	Accounting and bookkeeping services	Advice on labour law, tax law and the legal situation of foreign companies
Firm K	Surveying services from the legal side	Services at the interface of geodesy and real estate management, e.g. preparation of geodetic documentation for the return of expropriated real estate, preparation of documentation for the purpose of regulating the legal status of real estate, preparation of division projects for court purposes, investment service

(continued)

Table 6.1 (continued)

Firm	Offered services	Detailed description of the services
Firm L	Scientific and research services	Research and implementation works, development of studies and expert opinions, as well as consulting and advisory services in the field of maritime economy
Firm M	Advertising services	Comprehensive services related to the functioning of the company in the field of creating a multifaceted image (e.g. e-commerce services, graphic design, printing, creating visual identification)
Firm N	Geological, geotechnical and geohydrological services	Preparation of various types of geological and engineering, geotechnical and hydrogeological documentation, as well as conducting research in this area, preparing expert opinions

Source: own study based on sources collected on the surveyed companies

- advertising services (3 companies);
- accounting and bookkeeping services (4 companies);
- research and development services (1 company);
- technical design services (2 companies);
- broadly understood IT services (2 companies);
- surveying and geological services (2 companies).

The largest part of the surveyed companies were companies offering accounting and bookkeeping services (4 companies) and advertising services (3 companies). The sample of 14 companies included 6 different types of activity.

Apart from the type of services offered, the respondents were also asked to specify whether the services offered by their companies were routine, standard or not. Neither company indicated offering only routine services, and some companies declared offering only the latter. Some of the surveyed companies (Company A, Company D, Company E, Company F, Company H, Company I, Company K, Company L, Company N) declare that they offer both routine and non-routine services—tailored to the client's needs (the so-called tailored measure), while the proportions of one to another are varied. On the basis of the respondents' answers, 5 code categories were created: predominance of routine services, advantage of non-routine services, standard service with the possibility of modification, standard service in a non-standard environment and non-routine services. Selected quotations from the respondents' statements about the routine and non-routine nature of services are presented in Table 6.2.

For example, according to the owner of Company A, the majority of services are routine, repetitive services, mainly accounting and bookkeeping services. Conversely, non-routine services are related to advisory activities in the legal and tax area. Various cases and problems of clients force this type of activity to think outside the box and look for creative solutions. The president of Company H also pointed to the advantage of routine services, which constitute two out of three product groups.

Table 6.2 Routine and non-routine nature of the services offered by the surveyed KIBS companies

Code	Quotation
The domination of routine services	"A large proportion of [services] are routine, repetitive. Mainly in the field of keeping tax and accounting books. However, in the field of legal and tax consultancy, here each case is new and causes that it constantly opens new thinking, a new way of acting, inventing, etc. Here there is more creativity, while in such bookkeeping, i.e. the activity of an accounting office, there are repeatable things". [Firm A]
	"We have three product groups, two product groups provide repeatable services, and one typical made-to-order product group". [Firm H]
The domination of non-routine services	"*[Services are]* very little repetitive, basically only accounting, which is one fifth or one sixth of our work, is routinized. But we can't passively post what the customer brings; we need to guide him and let him know what he needs to do to make the project a success". [Firm F]
Standard service, modifiable	"There is a standard product and it can be implemented in some standard way, but in a situation where the customer has specific requirements or you need to integrate with external systems, or customers have some special requests, we adapt this software to the specific needs of individual customers". [Firm D]
Standard service in a customized environment	"They *[services]* at some level they are completely trivial and standardized. (. . .) But the reality is very different from this assumption, because the product cannot be standardized here. Each time something comes out completely, so to speak, 'out-of-the-blue', stories can come out that cannot be predicted and cannot be implemented". [Firm K]
Non-routine services	"It is known that most do not need the same, so depending on the type of company and work performed there, each offer must be tailored to the needs of a particular company. (...) We, as a company, try to meet the needs of the market". [Firm G]
	"Creative activities are unique activities, because each client is different, has different needs, and is also different, as if it brings a different human factor. These are completely different people, different emotions, a different task, so it's different every time. Maybe we would like to rest sometimes, have more template clients, but this is not possible". [Firm M]
	"Basically when is it explained what it is *software house*,[a] because I run such a company, we are talking about a tailor who sews a jacket or tailored clothing, so that we adjust to the client every time. Here it is worth adding that each project is really different because, for example, the knowledge we touch, the field we come into contact with, is always different, different project methodology, different client, etc.". [Firm C]

Source: Own elaboration
[a]*Software House* is a software development company

In turn, the respondent from Company F pointed to a clear advantage of non-routine services, which, in her opinion, constitute four fifths or even five sixths of all offered services. As customer reporting situations are usually non-routine, they require a differentiated approach. Sometimes, as in the case of Company K, the services are by definition fairly standard, but often there are unforeseen or unusual problems that need to be resolved. In other words, the service provided theoretically includes standard elements (e.g. preparation of division projects for judicial purposes), but when analysing the client's situation, it turns out that some documents are missing or the actual state is different than presented by the client. Then the service cannot be delivered in a standard way, but there is a need for its modification and an unconventional approach.

From time to time, KIBS offers a standard product, but it can be modified according to the customer's requirements. This is the case for Company D, which offers sales support software. Customers often have other solutions that must be integrated with this software, or they need specific functionalities that are not present in the delivered software, and then they should be adapted strictly to the customer's needs.

Some companies declared to offer only non-routine services tailored to a specific client. These firms include Firm B, Firm C, Firm G, Firm J, and Firm M. Offering non-routine services may require customers whose needs vary and therefore it is impossible to develop a single, standardized service for them. This was the case, for example, in the case of G Company, which meets the requirements of customers and offers services tailored to their needs. The owner of Firm M pointed out a similar aspect, according to which the service must be non-standard because it is based on creative activities and responds to various customer needs. At the same time, a respondent from Company M pointed to the difficulty of offering this type of service—they make the company constantly work on new solutions that are more demanding than the standard ones. In turn, the owner of the C Company referred to the analogy of a tailor sewing tailored clothes, each time creating a new, tailored product. The necessity to create a non-routine service results, among others, from the diversity of the areas the project is to concern, as well as the diversity of clients and knowledge that is required to provide a given service.

Although many researchers indicate that KIBS are non-routine and the level of their adaptation to customer needs is higher than in the case of other services (Corrocher et al., 2009), and Muller and Doloreux (2009) stated, when defining these services, that they are non-routine, they may not be. The surveyed companies often declared that they offered both non-routine and routine services. Most of the surveyed companies indicated both types of services, depending on the client's needs, and two companies even stated that they mainly offer routine services (Company A and Company H).

Offering routine services allows KIBS companies to achieve many potential benefits. First, it reduces the risk associated with offering only non-standard services. Second, if a company offers routine services, this saves time and resources on providing them. The preparation and delivery of a routine service is usually less time and labour consuming than for a non-routine service. Third, by offering standard, routine services, it is possible to create a modular service consisting of the individual elements required by the client.

The advantage of routine or non-routine services may be related to the market and the demand for a given type of service. KIBS SMBs may be looking for a niche and it will either be tied to a standard product that is required by other companies or it will require offering custom services. This may be a factor that determines the type of service offered. Some companies specialize in offering a specific service (e.g. implementation of X software) and this is what they mainly do, even if the client requires an unconventional approach from them. They do not need to expand the spectrum of their activities, because they do well in the market and have a product that is sought after by customers. In turn, other KIBS companies may offer products that are not available on the market, and which are manufactured to the order of a given customer—then each service will be non-standard and will be created in response to a specific customer need.

Standardization of the service may also result from the level of creativity of the created solution. Some KIBS will require a high level of creativity (e.g. design services or advertising services), while others—much less (e.g. accounting and geological services). The greater the level of creativity required in developing a KIBS service, the more custom the service will be (Bettiol et al., 2015). As already mentioned, it can be assumed that the level of non-standard service will determine the intensity of knowledge flows (all kinds).

6.1.2 Organizational Structure

The next question was about the structure of the organization. One might expect the structure of the surveyed organizations to be flat, as this supports collaboration, teamwork and knowledge flows. Moreover, the flat structure is typical for small entities (Cardon & Stevens, 2004). It turns out, however, that in some of the surveyed entities the structure is hierarchical or mixed. The obtained responses were classified into four code categories: hierarchical structure, flat structure, flat-hierarchical structure or other structure. The organizational structures indicated by the respondents together with the related quotations of the respondents' statements are presented in Table 6.3.

Some of the surveyed companies stated that their organizational structure is hierarchical (Table 6.3). For example, the owner of Company A stated that the organizational structure is hierarchical, although the owner tries to create friendly relationships in which employees feel that they are members of the "company family". The necessity of a hierarchical structure in the case of Company A results, among others, from responsibility for the decisions and actions taken, the consequences of which could be severe for the organization. The situation is similar in the case of Company K, but the respondent from this company also noted different levels of employee knowledge—not all of them are able to properly complete the task or make difficult decisions. Company N and Company L also declared having a hierarchical structure.

Table 6.3 The structure of the surveyed KIBS companies

Code	Quotation
Hierarchical structure	"The structure is hierarchical, although we try, regardless of this hierarchy, a bit like the American, to create a family, that is, to be in such friendly and collegial relationships. This structure results from responsibility and sometimes also decision-making". [Firm A]
	"*[Structure is]* hierarchical, definitely because there are so many people who are able to make a decision. These are not the five employees who know little. Me and another boy, also my father, who helps [we make decisions] sometimes, and it's definitely a hierarchical structure". [Firm K]
Flat structure	"Flat, definitely flat, even the CEO is with everyone on you". [Firm J]
	"There is definitely a decision president, but I decide about smaller problems, and he doesn't even like too many questions. So he only wants to be on those property rights and liability for which the law firm is insured. And I have to solve minor problems, and the other seven accountants are treated almost equally, each of them has its share in solving everyday problems". [Firm F]
Flat-hierarchical structure	"There is a management board, which is really me. I have specialist managers under me who already work directly with the team, but due to the fact that we are a small organization, I very often get involved in such topics, where I work directly with such people who perform such 'hard' work". [Firm C]
	"Formally, the structure is hierarchical, but in practice it is flat due to the fact that we are not a large company - there are 12 people here; despite the fact that there is this hierarchical organizational structure somewhere, it is flattened in the practice of such everyday work". [Firm D]
Other structure	"We have a very strange structure, which means most people find it strange. There is my humble person, there is X who manages the office, there are three product group managers below me and there is a resource pool under the product group managers. Project managers and engineers with different qualifications "swim" in the resource pool. When they start a new project, they select a project manager for a new project, and the project manager selects engineers from those who are currently working on their own development and are not involved in any commercial products, so it is difficult to clearly define who is whose boss sometimes, because these resources flow freely between departments. (...) Self-organizing organization". [Firm H]

Source: Own elaboration

Some companies indicated a flat structure, e.g. Company B, Company F, Company G, Company I and Company J. As the manager of Company F stated, apart from two people who make strategic decisions (her and the president), the other seven (accountants) employees are treated equally and each of them has similar competences and responsibilities. It is true that two of them, due to their more complete education and experience, share their specialist knowledge with the rest, but this does not affect the structure of the organization.

An interesting example is Company H, which has an unusual structure. It can be concluded that it is partly a hierarchical and partly a process structure. The president of the board of this company used the phrase "resource pool" to illustrate the availability and flow of human resources in individual projects. Part of the structure of this company is hierarchical (president of the board, office manager, three product group managers), and some are organized in a process (pool of resources that are selected and allocated to the implementation of individual projects). This mixed structure was described by the respondent as a “self-organizing organization”. Importantly, he found that in the past the organization had a flat structure, but this formula was not effective with 40 employees. Company H is slowly approaching the size of a medium-sized company, which may affect processes and changes in both the structure and nature of work (e.g. difficulties in supervision, division of duties, assignment of tasks, etc.).

In turn, Company C, according to the owner, has a flat-hierarchical structure in the sense that the owner of this company manages managers—specialists, but because it is a small organization, the boundaries of subordinate in some aspects blur and the owner also takes part in the implementation of ordinary tasks. The situation is similar in Company D, where the structure is formally hierarchical, but due to the fact that 12 people work there, it is flattened in everyday work. The hierarchical-flat structure was also indicated in Company E, where programmers have full autonomy in the area of a given solution they use, but some solutions are imposed by their superiors. Therefore, the respondent from this company described the structure of the organization as mixed (hierarchical and flat).

To sum up, the examined organizations, contrary to expectations, are characterized by a structure not only flat, but also hierarchical or mixed. Although a flat structure promotes the exchange of knowledge and cooperation between employees, the surveyed companies sometimes choose a hierarchical structure, or at least formally have one. In one case, the structure was referred to as hierarchical-process structure, because it contains elements of a hierarchy, as well as free selection of resources for the implementation of individual projects.

Based on the presented research results, it can be concluded that the surveyed KIBS companies make the type of their structure, to some extent, dependent on the level of employees' knowledge and responsibility for the actions taken. In some cases, if employees have similar knowledge resources or their knowledge concerns different (disjointed) areas, the organization structure is flat, and employees have a lot of autonomy in performing their work and making decisions. In other cases, if the actions or decisions taken may affect the company's situation (e.g. cause legal problems), the structure of the organization is at least partially hierarchical and key decisions belong to the people with the most knowledge or authority (in the case of regulated industries),[1] e.g. legal, accounting or geodetic.

[1]Regulated industries are largely dependent on widely understood legal acts.

6.2 Knowledge in Companies Offering KIBS

This part of the work presents the perception of knowledge by the respondents from KIBS companies as a source of competitive advantage, the types of knowledge most important to these companies and the reasons for its importance. These issues form the basis for further considerations on the distribution of knowledge in KIBS companies, as well as knowledge related risks and ways of eliminating them. To fully illustrate the flow of knowledge in organizations, it is necessary to analyse the knowledge itself and its overall perception by the respondents.

6.2.1 Knowledge Essential for the Functioning of KIBS Companies and as a Source of Competitive Advantage

KIBS sector companies are aware of the importance of knowledge for their functioning. All surveyed organizations, when asked whether knowledge is important for their functioning, replied in the affirmative, often adding that even very much. Many companies indicate that the most important type of knowledge for them is professional knowledge needed to provide the service. Some of the surveyed companies even indicate that knowledge about customers is not as important as professional knowledge, because customers themselves go to a given company if it provides its services well. The price criterion is often not important for customers, the most important for them are professionalism and the ability to provide the service (quality is the criterion for choosing a service). Some clients depart from competing companies because they have not received an adequate level of service there and decide to use the services offered by the surveyed companies. Each of the surveyed companies indicates a different type of specialist/substantive knowledge that is important to them, depending on the type of services they offer. Descriptions of individual types of knowledge relevant to respondents' companies are presented in Table 6.4.

The reasons for the high importance of professional knowledge for the activities of his company were explained by the respondent from Company A. In his opinion, knowledge is the basis for providing a service, and experience helps in its implementation. Without adequate knowledge, supported by experience, the functioning of this type of organization on the market would not be possible in the long term. Also the owner of Company D emphasized that knowledge and knowledge-based solutions are in fact the main added value that the company can offer to the client. Having established expertise is very important because of the potential ramifications of not having it. As the respondent from Company J stated, the lack of appropriate knowledge is primarily a problem for the client who will suffer the consequences of receiving an unreliable service, making wrong decisions or actions. In general, the respondents indicated the following reasons for the importance of knowledge for the functioning of their organization: knowledge is the basis for the provision of a service, and the consequences of the lack of knowledge can be very severe. Other types of relevant knowledge, according to respondents, are knowledge from

Table 6.4 The most important knowledge for the functioning of the surveyed companies KIBS

Firm A	Offered services	The most important knowledge for the functioning of the organization
Firm B	Tax advisory services and accounting and accounting services	Legal and tax knowledge as well as experience
Firm C	Advertising services	Production expertise
Firm D	Services for the production of dedicated software	Knowledge of technology, methodology and knowledge about the business and the client's enterprise
Firm E	Software development and implementation services	IT knowledge regarding the services provided
Firm F	Advertising services	Specialist knowledge of website design (e.g. user experience[a])
Firm G	Accounting and accounting and management services	Knowledge in the area of legal and accounting regulations
Firm H	Technical design services	Professional knowledge
Firm I	Technical design services	Any knowledge
Firm J	Accounting services	Knowledge of tax regulations
Firm K	Accounting and bookkeeping services	Knowledge in the field of tax law
Firm L	Geodetic services (the legal aspect)	Specialist knowledge accumulated over the years and experience
Firm M	Geological, geotechnical and geohydrological services	Technical knowledge in the substantive scope covering the activities of the organization
Firm N	Advertising services	Knowledge in the area of IT, psychology, various programs, applications related to design and programming
Firm A	Geological, geotechnical and geohydrological services	Specialist knowledge in the field of the company

Source: Own elaboration

[a]User experience is a term used in the context of software or websites and referring to the entire experience that a user experiences when using an interactive product

experience, knowledge about competitors, knowledge about customers and knowledge about current activities. Detailed descriptions of individual categories using quotes from the respondents' statements are presented in Table 6.5.

In addition to having professional/specialist knowledge in a given area, some respondents also emphasized the importance of knowledge derived from their own experience. According to the respondent from Company C, lessons can be learned from each project, and the most important thing is to understand the errors and shortcomings so that they can be avoided in the future. This type of procedure can be called an AAR (After Action Review) technique, which means a structured review or summarizing process to analyse what happened, why it happened and how it could be better done in the future by participants and project or event managers (Schindler & Eppler, 2003). The technique comes from the US military, where it was developed and introduced.

Table 6.5 Reasons for the importance of knowledge for the operation of KIBS companies and types of relevant knowledge

Code	Quotation
Knowledge as the basis for the provision of a service	"In our case, knowledge is basically the basic basis for providing services". [Company A]
	"Besides, our knowledge is our main added value that we can actually offer the client. Knowledge and these knowledge-based solutions". [Company D]
	"We must have this knowledge very well established and have it at all". [Company J]
Consequences of lack of knowledge	"If not for solid knowledge and experience, of course, also, then this type of entity in our conditions would not be able to survive for a long time. That is, either he would have to go down in his activity, or he would stop working at all, because there are so many risks that here, without this solid knowledge, it is impossible to work". [Company A]
	"Unfortunately, ignorance in this area has very big consequences for the customer". [Company J]
Knowledge derived from experience	"The most important thing is the knowledge that comes from experience. It never happens that a project will turn out perfectly and very well. In every project, we fail to a greater or lesser extent. If the failure is small, then the project is successful. We say this because we have earned some money and managed to implement it, *etc.* And now the key lies in avoiding the mistakes that were made earlier during the next project/project. And here we are talking about such knowledge resulting from experience". [Company C]
Knowledge about competition	"Not every customer is for us. There are various factors that make a client decide to work with us and not with another company. So I'm not so concerned about this competition. I am more focused on how my company can grow, I am not looking at the competition". [Company M]
Knowledge about customers	"Knowing what industries customers come from and what products they need". [Company B]
Knowledge about current activities	"The second important type of knowledge is knowledge about even the current functioning of the organization, for example". [Company L]

Source: Own elaboration

Some companies have found that knowing what their competitors are doing is not important to them (e.g. Company K, Company M). According to the respondent from Company M, developing one's own company is more important than observing the competition, because when choosing a company for cooperation, the client is guided by various criteria and it makes no sense to focus on all of them.

When asked about the perception of knowledge as a source of competitive advantage, representatives of all surveyed companies agreed with the statement that in their case knowledge is the source of their competitive advantage, and various justifications or explanations for this state of affairs were given. Some companies stressed that they offer better, more complete services that include additional knowledge that competing organizations do not. This opinion was expressed by the respondents from Company A, Company F or Company J. In their opinion, it is precisely better, more complete knowledge that allows these organizations to achieve better results and gain a competitive advantage. The respondent from Company J emphasized the lack of competing in terms of price with other companies, pointing to the importance of a well-performed service, based on solid knowledge and skills. The owner of Company F was of a similar opinion and also presented arguments for not competing with the price. Even if customers refer to other companies offering cheaper services, over time they appreciate Company F's wide range of services and are willing to pay more for its service.

Also the respondent of Company I indicates that greater knowledge of various regulations from various fields allows for gaining a better competitive position. It is not only about professional knowledge, but also knowledge related to the specificity of the operation of companies—clients, e.g. construction companies, transport companies or companies that deal with exports. Selected fragments of respondents' statements concerning the perception of knowledge as a source of competitive advantage in the surveyed companies are presented in Table 6.6.

Another important aspect of knowledge was pointed out by the technical director of Company N. In his opinion, having appropriate knowledge allows for faster performance of tasks or reduction of costs of their implementation. The owner of Company B was of a similar opinion, according to which knowledge enables us to be competitive in the sense that it allows us to offer more modern products at a lower price. Knowledge can therefore influence a company's competitive advantage, helping it to gain a cost advantage or an advantage resulting from innovation. Thus, it is not only a direct source of competitive advantage (the company has more knowledge than its competitors, and therefore customers choose it), but it can also indirectly affect other methods of gaining an advantage (e.g. resulting from lower costs or better quality products).

According to the owner of Company K, the unique combination of knowledge and experience allows the company to function in its niche, and the difficulty in codifying this knowledge and transferring it gives the company a competitive advantage that is difficult to take over. On the one hand, codifying knowledge may bring many benefits (e.g. the possibility of using knowledge in the future by a larger number of employees, etc.), on the other hand, the lack of codification of knowledge may allow KIBS companies to maintain a competitive advantage. This is one of the potential reasons why KIBS companies may be less willing to use codification and knowledge distribution tools. A different approach was presented by a respondent from Company L. According to her, knowledge is a source of competitive advantage in the sense that the company has employees that someone else does. It is the right staff with knowledge resources that is the source of competitive advantage.

In turn, according to the owner of Company H, the possessed knowledge does not currently (and will not be significant in the near future) of much importance, and what matters is only how much a given company can learn in a short time. In the opinion of this respondent, the key is gaining new knowledge, and protecting the

Table 6.6 Knowledge as a source of competitive advantage in the surveyed KIBS companies

Code	Quotation
Better knowledge of the characteristics of customers and their environment	"Greater knowledge of different regulations from different fields; when different groups of customers come, or, for example, construction companies or transport companies, or companies that deal with exports or work with ships, with a shipyard, for example, depending on what such companies do, you need to know the relevant regulations". [Company I]
Greater knowledge resources	"Our company has never competed on the market with price, but mainly with the knowledge and skills that we have and which we have acquired, say, for 13 years. Our services are not the cheapest on the market, although I think that the client is able to pay for a well-performed service, a professionally performed service. And he should pay. She should be included. (...) Clients from them [other companies] come to us, because there is no such knowledge, this vast amount of knowledge. We will respond to the client only when we are 100% sure that what we say is supported by regulations, and if we do not know a certain aspect, or maybe not that we do not know, but are not entirely sure, we correspond with the office, we read the recipes again and try to put it together and interpret it. So this is what our company generally competes with". [Company J]
Faster performance, lower costs, better products	"Knowledge allows you to do certain things faster or at lower costs". [Company N]
	"Knowledge is a source of competitive advantage in the sense that we can be competitive because we produce more modern things cheaper, because we know how to use innovation". [Company B]
Knowledge difficult to pass on	"On the one hand, the knowledge of the current regulations, and on the other, this historical knowledge is extremely important, because in order to evaluate the materials on the basis of which we are to organize the work, you need to know where they came from. It is unwritten knowledge related to acquired experience, so it is difficult to transfer and thanks to that we have our niche". [Company K]
Employee knowledge	"We have staff that no one else has". [Company L]
The speed of acquiring new knowledge and applying it, continuous learning	"Currently and in the coming years, the knowledge possessed does not *[and will not]* matter much. In fact, the most important thing is how much we are able to learn in a short time. (...) I think that the most important thing is to acquire new knowledge, and protecting the already possessed is not so important". [Company H]

(continued)

Table 6.6 (continued)

Code	Quotation
	"Knowledge is very important, but maybe not so much constant knowledge as improvement. Continuous work, because it is both this e-commerce and these creative activities, these are areas where if you do not develop, it will be lost in a moment, so it is very important". [Company M]
	"Learning about new technologies and having the courage to apply them gives you that *[competitive]* advantage, but it also comes with risks. That is, you never know which technology will catch the 'wind in the sails'. (...) Generally, today, when you create software, you use many technologies, and I would say that using this knowledge, gathering this knowledge, reading what appears, reading first opinions, *etc.*, and then having the courage to apply it, knowing the fact that you can fail in this topic is actually a huge advantage". [Company C]
	"Knowing about the latest trends, but also knowing what trends to avoid *[is important]*. If you know that some solutions should not be introduced because they are only temporarily a fashion or look cool, but are problematic in terms of use, we will not introduce this solution. So this is definitely a key advantage. If we are able to explain it to the client and he is able to understand it, I think he will be happy to use our services". [Company E]

Source: Own elaboration

possessed knowledge is irrelevant. This shows the speed of ageing or obsolescence of knowledge. The owner of Company M shares a similar opinion, according to which it is important not so much to have knowledge as to improve. The type of business carried out by Company M necessitates continuous development and the acquisition of new knowledge, or else you may quickly go out of business.

On the one hand, having and applying new knowledge gives you a potential advantage, but these aspects also come with risks. The owner of Company C points to this issue. Acquiring new knowledge is very important, but it should not be used without reflection; it is necessary to verify it and treat it with distance. It may turn out that new knowledge does not work in every environment or that its implementation in the long run is problematic. On the one hand, sometimes it pays off to take a risk and apply a new solution, and on the other hand, it is worth considering it carefully and being aware of potential difficulties. The owner of the E Company also points to the need for a good understanding of new knowledge and solutions and the ability to communicate them to the client as factors contributing to a competitive advantage.

6.2.2 Distribution of Knowledge in KIBS Companies

In the KIBS companies surveyed, knowledge was not usually distributed evenly—only one company stated otherwise (Company I), although some organizations declared that they tried to distribute the knowledge evenly in them. Uneven distribution, according to many respondents, resulted from the specialization and different levels of knowledge possessed by employees. The issue of knowledge distribution along with the reasons for uneven distribution of knowledge and the solutions to this problem are presented in the respondents' opinion in Table 6.7.

An example of a company that tries to implement solutions enabling an even distribution of knowledge is Company H. The company planned to introduce a DMS system, which was to be used to distribute knowledge relevant to employees (e.g. regarding regulations, holidays or applied standards), and had already introduced the company's social network called Yammer. The situation was similar in Company F, where the owner tried to distribute the knowledge evenly, because when one employee does not show up at work, others have to replace him and serve the customer. While some employees specialize in certain areas, all of them need to have some knowledge related to the services they provide.

Some kind of knowledge can and must be evenly distributed, while others cannot. This aspect was pointed out by a respondent from Company D. In his opinion, if the entire team is working on some tasks, the knowledge about these tasks must be evenly distributed, e.g. in the case of the implementation team. In turn, if a given employee deals with a narrow area and only he performs tasks related to it, then, according to the owner of Company D, distribution of this type of knowledge is not necessary. A similar opinion was expressed by a respondent from Company L, according to which general knowledge concerning all employees is passed on to them, and specialist knowledge of individual employees is not. A similar opinion was presented by the technical director from N Company, claiming that knowledge cannot be evenly distributed.

Unequal distribution of knowledge is, according to the respondent from Company G, related to the ability of individual employees to assimilate knowledge: some will be able to acquire more knowledge, others—smaller; some employees will show greater initiative in this area, while others—less. In turn, the respondent from Company J indicated differences in the distribution of knowledge related to the length of service in the company. According to her, the knowledge of an employee who has just been hired will never be equal to that of an employee with 10 years of experience. This is related to the experience gained while solving various types of problems and acquiring specific knowledge related to work in a given organization. Knowledge is not evenly distributed also in Company K, because individual employees are characterized by different levels of knowledge (two employees have much more knowledge than others), and in Company B, Company C and Company E, different employees have different types of deeper knowledge in the areas in which they specialize. The situation is similar in Company M—according to the owner of this company, in the industry in which the company operates, there are

Table 6.7 Distribution of knowledge in the surveyed KIBS companies

Code	Quotation
Uneven distribution	"I think there are some employees who are much more knowledgeable. It is never the case that everyone has the same knowledge. There are also people who know more about individual work segments. So it cannot be said that there is an even distribution of knowledge". [Company B]
	"It is definitely not even. We have all kinds of knowledge. I can say that one is closer to this ideal of evenness and spread, which does not mean that it is close. It is only closer, because it can be far away all the time". [Company C]
	"Certainly uneven, because employees have different areas in which they specialize. Even if, let's say, they come to work and have some general knowledge about something". [Company E]
	"The distribution of knowledge is not even, and that's the problem". [Company H]
	"I don't think everyone has the same knowledge, because everything is developing so much that we work with people who specialize in some areas". [Company M]
Uneven distribution—causes	"Urgent is the enemy of the important. I know that it is important that knowledge is, that knowledge management is done properly, guarding it, *etc.* The problem is that it is never urgent". [Company C]
	"Everyone will get as much as they can assimilate. I mean, there aren't any chapters of secret knowledge so as not to share something with someone. I really like when employees gather together and obtain this knowledge themselves or are interested in what they need". [Company G]
	"Young employees will not have the same knowledge base as employees with 10 years of experience. And I don't think it can be achieved in any company. On the other hand, if we were to put two employees with 10 years of experience and two employees with 5 years of experience, their level of knowledge is more or less the same". [Company J]
	"For example, I know something about positioning, about programming, but I wouldn't do it myself. It is an industry that there are narrow specializations, and then, when more complex projects are done, it is nice to share this knowledge for the common good, because we will all earn money on it, the client will be satisfied, then he will come back to us". [Company M]
	"Knowledge cannot be [distributed] evenly because it is assigned to specific topics. I don't know everything about everything either, but I know roughly about each project. How is it realized, when it starts, when it ends. On the other hand, people who document these projects or implement them know this one topic inside out, that is, they have 100% knowledge on its what happened in the given area is going on, but they don't necessarily know about other things that don't concern them". [Company N]
Knowledge that must be evenly distributed	"It depends on the specific scope. If there are such things, *e.g.* implementation issues, then there is a whole team of implementers that needs to know it, and then the knowledge is

(continued)

Table 6.7 (continued)

Code	Quotation
	passed on to everyone. And we have a few such employees who have a specialization there, and they mainly deal with this topic, so it is not very useful to share some aspects of this knowledge, because it would not contribute much". [Company D]
	"General matters that concern all employees - we have this information for employees on the site, so everyone can have a look, first of all. Secondly, if something concerns all employees, it is sent out by e-mail, there are ordinances, also the basic information that everyone should read, is sent to all employees". [Company L]
Uneven distribution—solutions	"We will implement DMS and DMS[a] it will also be a means of distributing knowledge. So all kinds of standards, regulations will be in one, centralized place, where they will be easy to find. Queries will be recognizable, it will be possible to look for it in many different ways, *etc.* After entering the keyword 'vacation' I will know everything about leave, how much leave I have, how much I am entitled to, why, what should I fill in *etc.* Second thing - we use Yammer. Yammer[b] also did such a knowledge centre a bit. On this Yammer people throw a lot of things that are happening there in the world, on the market, what appeared, why, who does what and so on". [Company H]

Source: Own elaboration

[a]DMS systems, i.e. Document Management System. They are used to complete, collect, index, share and search for various documents created within the company as well as those coming from outside

[b]Yammer is an enterprise social network that operates in the freemium model. It is used for internal communication in companies and organizations. It is an example of a social software used in a business environment, https://pl.wikipedia.org/wiki/Yammer, access date: 12/08/2018

narrow specializations and employees are not able to have knowledge of issues that they do not deal with on a daily basis.

Based on the opinions of respondents from the surveyed KIBS companies, it can be concluded that the distribution of knowledge depends, inter alia, on several factors:

- *the need (or lack thereof) for all employees to have similar knowledge:*[2] such a necessity often occurs in companies where all employees are involved in similar tasks (e.g. in accounting firms), while it is rare in companies where employees perform various tasks, each employee is responsible for a part of a larger project

[2]In the surveyed companies offering legal/accounting/advisory services, knowledge was passed on to each other by all employees, e.g. one person went to the training and then returned and trained the others or photocopied materials for them, etc.

or service element (e.g. in IT companies) and employees are not able to have the same knowledge;[3]

- *experience and internship of people working in the organization:* often, the greater the experience and seniority of an employee in a given organization, the more knowledge they have compared to a newly employed employee;
- *hierarchy or organizational structure and related decision-making responsibilities:* if decisions can be made only by the most experienced people or those with the highest formal powers, then even distribution of knowledge will not occur;
- *the kind of knowledge that is being distributed:* some kind of knowledge will be distributed more evenly (e.g. general knowledge related to the functioning of the company, rules, applicable to all employees), while another—less evenly (e.g. specialist knowledge related to the performance of tasks to which selected employees are assigned—others do not need to have this knowledge because they are not involved in the implementation of tasks).

After discussing issues related to knowledge distribution in KIBS companies, practices related to knowledge management in these companies will be presented. These practices indicate system solutions for knowledge management used in the surveyed organizations. They constitute a certain point of reference for the subject of knowledge flows in KIBS companies discussed in the next part of the work.

6.2.3 Practices Related to Knowledge Management in the Surveyed KIBS Companies

The surveyed companies were asked to indicate which practices related to knowledge management they introduced, but the wording "knowledge management" was not used so as not to influence the responses of the respondents. The solutions used in organizations were asked about and the respondents were presented with a list of practices. All doubts about solutions or the level of their implementation (e.g. a solution was introduced, but it was rarely used) were clarified on an ongoing basis in order to obtain answers as close to reality as possible.

The practices introduced by individual companies are presented in Table 6.8.

The only practice related to KM introduced by all surveyed companies is building and maintaining employees' knowledge and skills (through seminars, specialist training, mentoring, knowledge refresher courses). This is surprising since most of the surveyed companies are small organizations that are reluctant to train their employees for the fear of leaving them to competitors or large entities (Panagiotakopoulos, 2012). Investing in employee development by the surveyed

[3] In the surveyed IT companies, knowledge was often passed along with tasks; when a given thing had to be done, the employee received knowledge that he did not have and which he needed to perform the task.

Table 6.8 Practices related to knowledge management introduced in individual companies

	Company A	Company B	Company C	Company D	Company E	Company F	Company G	Company H	Company I	Company J	Company K	Company L	Company M	Company N	Summary
Acquiring and storing technical or market knowledge in an electronic repository or in the form of documents (manuals, descriptions)		X	X	X	X	X	X	X	X	X	X	X	X	X	13
Using e-mail to share and transfer technical and market knowledge		X	X	X	X	X	X	X	X	X	X	X	X	X	13
Using social media (wikis, blogs, Facebook applications) for publishing and obtaining information		X			X		X	X		X			X		6
Building and maintaining employees' knowledge and skills	X	X	X	X	X	X	X	X	X	X	X	X	X	X	14

(continued)

Table 6.8 (continued)

	Company A	Company B	Company C	Company D	Company E	Company F	Company G	Company H	Company I	Company J	Company K	Company L	Company M	Company N	Summary
(through seminars, specialist training, mentoring, knowledge refresher courses)															
Identification and dissemination of internal or external best practices regarding technological, market or operational solutions	X	X		X	X	X	X	X	X	X	X	X		X	12
Creating conditions conducive to the exchange of knowledge (availability of conference rooms, organization of rest areas	X	X		X	X	X	X	X		X	X	X	X	X	12

for employees)															
Rewarding employees (financial or non-financial) who share their knowledge	X	X		X	X	X	X	X	X	X		X		X	11
Organizing regular meetings (formal/ informal) to exchange information on projects, products, market etc.	X	X			X	X	X	X	X	X		X	X		10
Use of CRM or ERP software[a] not only to manage operational data about customers, but also to obtain information about markets,		X	X		X					X					4

(continued)

Table 6.8 (continued)

	Company A	Company B	Company C	Company D	Company E	Company F	Company G	Company H	Company I	Company J	Company K	Company L	Company M	Company N	Summary
ways of managing customer relations, etc.															
Use of the community of practice to share knowledge		X		X	X	X	X		X	X	X		X	X	10
Summary	5	10	4	7	10	8	9	8	7	10	6	7	7	7	

Source: Own elaboration

[a]ERP (enterprise resource planning), i.e. enterprise resource planning, is a method of effective planning of managing the entire enterprise resources

companies confirms the importance of employees' knowledge and skills for the functioning of KIBS companies. In addition to introducing this practice, many companies commented on its choice, stating that it is one of the most important. The necessity to use it results, according to the respondents, from constantly changing legal regulations and conditions, ageing knowledge that needs to be renewed or acquired new, or changing market and customer requirements

Other very common practices also include acquiring and storing technical knowledge about the market in an electronic repository or in the form of documents and using e-mail to share and transfer technical and market knowledge. These practices were introduced by 13 out of 14 surveyed companies. Less popular were the identification and dissemination of internal or external best practices regarding technological, market or operational solutions and the creation of conditions conducive to the exchange of knowledge (availability of conference rooms, organizing rest areas for employees) and organizing regular meetings (formal/informal) to exchange information on projects, products, market.

Eleven out of 14 surveyed companies also introduced rewarding (financial or non-financial) employees who share their knowledge, while, according to respondents, more emphasis was usually placed on non-material reward (e.g. through recognition, verbal or e-mail praise) than on material (e.g. bonus). Material rewards appeared less frequently, and even if they were trifles (e.g. a Star Wars mug or calendar at Company J). Some of the respondents who introduced rewarding employees indicated that it was related not only to the sharing of knowledge, but also to the knowledge possessed by a given employee (the more knowledge they had, the higher their remuneration was, but also the expectations related to sharing knowledge with colleagues or training them). Please note that financial rewarding of employees for knowledge sharing is difficult to organize (how can the level of knowledge sharing be measured, especially at an informal/non-codified level?). Moreover, companies from the SME sector often have limited financial resources (Torrès & Julien, 2005) which they cannot spend on rewarding employees for sharing knowledge. Finally, in KIBS companies the necessity to share knowledge seems natural, without which it is impossible to create a service for the client.

The surveyed organizations relatively often (10 out of 14 companies) also indicated using the community of practice to share knowledge, although for some of them it was necessary to clarify the term. Respondents indicated that they benefited from the support of both internal communities of practice within their organization and external ones (e.g. associations, online forums, professional colleagues, etc.). The president of the board of Company H pointed out an interesting phenomenon. He tried to gather experts who have already retired and create an unofficial community of practitioners who would meet regularly in a restaurant and talk about their experiences, sharing their knowledge with younger staff. This idea could not be implemented because the retired employees did not want to share their knowledge with others and only carry out a possible order. This barrier turned out to be insurmountable and the idea of such meetings was abandoned.

"We tried to create such a lodge some time ago. We tried to get people who have retired. We wanted to invite them once a week, order some food for them and sit them down in the downstairs bar, eat dinner and at least remember to leave the house. And at the same time as they would be sitting over a cup of coffee and eating dinner, we wanted to send people to them to solve problems, the more technically advanced ones. You would have to teach these people to talk, and that was the biggest problem. They were from a different system. They were brought up in a completely different way and I think that talking while eating makes them embarrassed. Complete turnout. They had the approach "I don't want to tell and if you have a problem, pay me and I'll do it for you"". [Company H].

The least popular practices included the use of social media (wikis, blogs, Facebook applications) to publish and obtain information, and the use of CRM or ERP software not only to manage operational data about customers, but also to obtain information about markets, ways of managing relationships with customers, etc. The first solution was introduced by 6 companies out of 14 respondents, and the second—by only 4.

When it comes to not using social media to publish and obtain information, the reasons for this are varied. One of the main reasons given was that there was no need to use them. Some companies (such as Company G and Company K) indicated that they are aware that they should use social media, but that they do not. This was often due to a lack of time regularly needed to create and promote news and material in these media. Company H, which uses social media (LinkedIn and YouTube; it was also planned to use Twitter), found that such services could attract the interest of partners and customers. Their disadvantage, however, is the need to create high-quality materials, which is associated with costs not only financial,

A very small proportion of the surveyed companies introduced CRM or ERP software. One of the companies that did this (Company J) has its own solution in this regard, because there was no solution on the market that would suit its needs and would offer the possibility of working with all data. In turn, Company A implemented this type of solution, but it turned out that it was not used. According to the owner of Company A, this was due to the lack of need and a relatively small number of employees who could cope with tasks without using this solution. The owner of Company M stated that although they do not have CRM yet, they must introduce it, because its lack makes their functioning difficult and it is impossible to live without such a solution.

In turn, the president of the management board of Company H stated that the environment in which he works is not yet ready for a CRM solution and in his opinion the introduction of such a solution would be a failure. This respondent also did not see the need to introduce an ERP solution, as in his opinion it is effective when employing at least 150 employees. In the case of the respondent's company, the white board and the visualization of the project flow solve the issues of the need for an ERP system and are a much cheaper solution. If such a system were needed, it would have to be created for the needs of Company H due to its specificity and mismatch with the specificity of solutions available on the market. The lack of CRM or ERP solutions in other surveyed companies was most often caused by the lack of the need to introduce them (e.g. Company A, Company D or Company F). The lack of need, in turn, could result from the size of the analysed companies (mainly micro or small entities), which did not necessitate the need to manage a large number of resources or customers.

The use of individual practices depends on various factors, but in general, it results from the current need in the area of knowledge of the surveyed companies. Some practices are natural for most organizations because they fit in with the way of doing business (e.g. acquiring and storing knowledge in an electronic repository or in the form of documents or building and maintaining employees' knowledge and skills). Knowledge becomes outdated so quickly that it needs to be refreshed and supplemented, both in repositories and in documents or in the minds of employees. These practices emphasize the importance of both codified and tacit knowledge possessed by employees. Conversely, if the company sees the legitimacy of using social media, it uses their possibilities, and if it is not useful, it does not. The introduction of KM practices is also partly related to the environment of KIBS companies and its characteristics. For example, according to the manager of Company J, the entities with which the company works largely use social profiles and electronic solutions, so, naturally, Company J also uses these types of tools. The respondent from Company J gives the following justification for this situation:

> "In general, sometimes you can learn more electronically than when it comes to face-to-face talk to the clerk. At least the information is given "in writing", and a person can always refer to it". [Company J].

These results were also confirmed in the author's study, in which the approach of small KIBS companies to KM was defined as an emerging approach, defined as: "an approach in which the practices, tools and methods of knowledge management come from the daily practices and learning processes of the company's employees. In fact, employees develop their own methods of learning, storing, searching and sharing knowledge in relation to real needs and practical problems to be solved. These methods and tools, which prove to be effective, useful and/or in line with everyday business practice, will later be developed to become established practices, and ultimately can be considered as the company's approach to knowledge management" (Zieba et al., 2016, p. 295). An important aspect of choosing the KM practice is, in the case of the surveyed companies, a real need for its introduction, often identified at the employee level. If a solution is not needed, no one tries to implement it or check its usefulness by trial and error. This part of the work presents aspects related to knowledge in the surveyed companies offering knowledge-intensive business services. First, the most important types of knowledge from the point of view of the functioning of these organizations are presented, the reasons for the importance of knowledge and the ways in which this knowledge influences the competitive advantage in these organizations. Then, the distribution of knowledge and the risks associated with knowledge were discussed, along with the methods of their elimination indicated by the respondents. Finally, the knowledge management practices used in the surveyed companies are presented. This part of the research allowed to present the full picture of knowledge and its perception by the respondents.

6.3 Knowledge Flows in Companies Offering KIBS

In the second part of the study, respondents were asked questions about the flow of knowledge. They were asked to present the perception of knowledge flow, to identify the types of knowledge flows that exist, to characterize the knowledge flow between KIBS and the client along with the problems and determinants of the problem and to describe the relationship between knowledge flows and workflows. Next, questions were asked about stuck knowledge and ways of counteracting this phenomenon, technologies and tools used in knowledge flows, as well as factors influencing knowledge flows. The obtained results will be presented in this part of the work.

6.3.1 Knowledge Flow Perception

The flow of knowledge was described by the respondents in various ways. On the one hand, people involved in the flow were described, but also the object of the flow (a kind of special knowledge) or the way in which the flow took place. Individual respondents drew attention to other aspects of knowledge flows (Table 6.9), but in general their understanding and perceptions of knowledge flows were consistent with the definition presented in Chap. 1. An important aspect pointed out by the owner of Company C was the use of flows to disseminate knowledge. When any relevant knowledge appears, it should be passed on to all people who can make use of it or use it in the implementation of tasks. Collaborators who, having authority, can easily pass on relevant knowledge to others play a special role here. A similar opinion is shared by the respondent from Company F, according to which the flows of knowledge are common, which takes place every day. When solving everyday problems, knowledge is passed on to all employees. Knowledge that can be useful in the company is immediately passed on to everyone for use in the future. Both respondents emphasized the applicability of the knowledge made available as part of the flow—because this knowledge can be useful, it must be transferred and disseminated. The respondents' perception of the flow of knowledge is presented in Table 6.9 and illustrated with selected quotes from their statements.

The flow of knowledge may be associated with the sharing of knowledge within the organization, as well as with the search for knowledge outside. According to the

Table 6.9 The perception of the flow of knowledge by the respondents of the surveyed KIBS companies

Code	Quotation
Disseminating knowledge through its flow	"I understand it in such a way that this knowledge is, firstly, universal, that is, all those who should have it have it in their heads. The girl who deals with finances in my company, *i.e.* the person responsible for finances, does not need to know/have knowledge about technology. There is no such need here, but I would say that there is so-called role knowledge, resulting from the role someone plays in our organization. And when we talk about flow, we are talking about the fact that when somewhere out there, let's say in a project, a decision is made to apply this and that practice in this and that technology, I understand the flow of knowledge that in a fairly short moment all people who play a similar role in the organization have this knowledge. This is the first point, and the second - then apply to it. Now, as if this information must be disseminated. This can happen in two ways: The first, in my opinion, the better is that it spreads in a whispered way. It's just that, at some point, everyone knows about it, which means it's communicated as if by colleagues in the role, the best authority ever. And the second way is, of course, meeting and communicating and communicating that we have or have had such an experience, experienced such things and are rich in some knowledge, and therefore we will follow it". [Company C]
The universality of the flow of knowledge	"I keep opening channels of this knowledge all the time. Every day I am persuading everyone and asking and telling them what I am doing; I inquire and also take part in solving problems at every workplace. That's how we go to the centre almost every day and each one comes out from behind the desk and we solve problems, and when I figure something, a topic, I also tell them, and we even put notes on it in one place". [Company F]
Sharing knowledge and looking for it outside	"The fact that everyone shares their knowledge is looking for that knowledge outside". [Company B]
	"*[the flow of knowledge is]* above all the sharing of information and experiences". [Company D]
Exchange/transfer of information and knowledge	"Exchange of information from one area of the organization to another, between employees and between different levels". [Company E]
	"*[flow of knowledge]* goes from me to employees and between employees, the flow of information is just like the flow of knowledge". [Company I]
	"I understand the flow of knowledge in the organization that we have employees who, for example, have been working in the company for 10 years, and during those 10 years they have already acquired a lot of knowledge. These are people like X, who feel very comfortable with international tax conventions and can analyse them. It is a person who knows well the provisions of lab or law, tax regulations, knows what

(continued)

Table 6.9 (continued)

Code	Quotation
	can, what cannot and such persons generally train persons that come fresh out there and have some stock of knowledge, but that knowledge is groundless. (...) If someone comes up with a recipe, someone finds that a change is being prepared, it is sent to everyone. And it works both for individuals and for business clients. So every change that someone spots, that he is getting ready (...), this is the moment when such information is passed on". [Company J]
	"We trade because we all do the same". [Company K]
	"*[the flow of knowledge]* is based on the fact that people who are better qualified in an open, friendly way, without prejudice, transfer knowledge during the implementation of some projects. Or it doesn't have to be better qualified people, because sometimes freshness also matters". [Company M]
The importance of conversation in knowledge flows	"Knowledge flows through conversations, by recommending some publications, e-books, websites, attending training courses, but I think that first, the first thing is a conversation, that if we don't know something, we consult each other and check, we drill down on the topic, we look for information beyond our environment. I mean, this is the basis, sort of digging into this environment around". [Company M]

Source: Own elaboration

respondents from Companies B and E, it is important to exchange knowledge within the organization, both between individual employees and levels, and in the case of a lack of internal knowledge—search for it outside the organization. The owner of Company K pointed out an important aspect of the flow of knowledge. In his opinion, the exchange of knowledge between employees is dictated by the fact that everyone in the company is doing the same thing. It can be assumed that if such a situation takes place and the knowledge of some people is potentially useful to others and the entire organization, the flows of knowledge are more intense. If there is a division of roles in the company and everyone is responsible only for part of the service, knowledge flows will be of a different type and will be related to the workflow (employee X did the service element X, employee Y did the service element Y, employee X does not have to and is not even able to have the same knowledge as employee Y and vice versa). An example of a first type of company is Company K, and a second type of company—Company C.

It was also pointed out that the flow of knowledge is naturally connected with the flow of information. This opinion was shared by the owner of Company I, according to which the flow of information takes place, as does the flow of knowledge, between him and the employees and between the employees themselves. A similar opinion was expressed by a respondent from Company J, who indicated both the case of transferring knowledge by more experienced employees to the less competent, and the mutual transfer of information related to the change in regulations. At the same time, the respondent emphasized the importance of openness towards people, because the lack of this openness causes reluctance to share knowledge and translates to a large extent into the atmosphere and quality of work. The owner of the Company D.

Communication and ordinary conversations between employees play an important role in the flow of knowledge. This aspect was pointed out by the respondent from Company M, according to which knowledge can be transferred in various ways, but the basic form is a conversation and consulting with other employees.

Summing up, it can be stated that the flow of knowledge in the surveyed KIBS organizations is understood very broadly. These are training, transfer of knowledge by some employees to others (mentoring, counselling), knowledge exchange between employees in the implementation of a task or transfer of knowledge vertically (from superiors to employees and vice versa) and horizontally (between employees). The respondents perceive the flow of knowledge as:

- exchange of information and/or knowledge: these terms are often used interchangeably and equated with each other, which may result from difficulties in separating information from knowledge and clearly stating when information becomes knowledge;
- sharing information/knowledge/experience: many respondents emphasized the importance of sharing knowledge or experiences in the flow of knowledge;
- internal employee training (older employees or more experienced employees train younger employees).

All these aspects of knowledge flows are described in the chapter that defines this phenomenon. Therefore, it can be concluded that the understanding of the term "knowledge flow" by the respondents is consistent with the definition proposed in Chap. 1. This is important because the lack of convergence in the area of the key term in the study could result in obtaining unreliable results.

6.3.2 Types of Knowledge Flows in the Surveyed Organizations

After examining how the respondents understand the knowledge flows, they were asked about the types of flows occurring in their organizations. The most common distinction between knowledge flows that occurred during the interviews was the division into internal and external flows. Some firms have also classified flows based on the type of knowledge being flowed. The types of knowledge flows indicated by the respondents and described in their words are presented in Table 6.10.

When it comes to internal flows in the surveyed companies, in some cases they seem to be a natural phenomenon. According to the owner of Company A, the flow of knowledge inside the company is strong because all employees are located in one place and together they try to solve emerging problems. The owner of Company B stated that knowledge flows in her organization are multidirectional, because individual employees have their own specialization, but also transfer their knowledge to

Table 6.10 Types of knowledge flows in the researched KIBS companies

Code	Quotation
Internal flows of knowledge	"The flow of knowledge inside is quite strong, *i.e.* it results from the natural conditions of working in a certain environment. That is, if six people are sitting in one room, someone is talking on the phone with an official or with someone and discusses a problem, and comes up with a problem, it immediately causes all who hear it to activate in the matter, and then, if they do not find a solution, then there is the greatest flow of knowledge and experience, because I imagine that as everyone has their own room or every cabin, it works and there is no this flow inside". [Company A]
	"At the level, it works in a multidirectional manner, because everyone has the opportunity to transfer their knowledge, everyone, individual employees have a specialization, but also pass this knowledge to other members. They are conversations". [Company B]
	"Knowledge exchange based on ordinary interaction - conversations, meetings in which we define further plans for the company's development. These are management meetings with employees, the entire development team". [Company E]
	"In the organization, it's mostly e-mails, teaching materials received somewhere at conferences or you know that sometimes I can't let the whole team go, I only let one person go, and that person collects materials, then we photocopy them and they are generally distributed. This is how it works between offices. Based on this information, we also create our studies, because some things are only available in x, so the girls explain it, prepare it, because not the entire team is x-linguistic, so they prepare such a study in Polish for other employees". [Company J]
	"Vertical, that is, I mean succession, that is, the seniors pass the knowledge to the younger ones, the younger pass them to the trainees, and this is how it flows through a waterfall, from top to bottom. And we also have horizontal transfer of knowledge, *i.e.* between employees at the same levels of the organization, who have to cooperate with each other, and it is obvious that not everyone is equally good at everything. This is where communication comes to our aid, and this is something we have put a lot of emphasis on for three years, just to teach people to talk and work in teams". [Company H]
External flows of knowledge	"The only knowledge that can be described in such a way to our environment is the knowledge about project management that reaches our client. Because we implement it and expect that he *[the client]* will follow the guidelines or possibly propose some changes, and this kind of knowledge emerges". [Company C]
	"Training, textbooks, this type of knowledge resources that we absorb". [Company E]
	"We deal with the external environment, in general, in professional matters related to strict design, we contact us, and we do not have such a need to open up widely". [Company G]

(continued)

Table 6.10 (continued)

Code	Quotation
	For our environment, *i.e.* potential customers or organizations that cooperate with us, we publish two magazines in which there is a huge amount of knowledge that we are very happy to share, and we send them. In our company, certain information can be provided free of charge, while the more complicated topics, more complicated studies, are paid for. If we can provide a study for free, we also provide such a study and these are studies in Polish, which are easy to understand, above all. It is known that whether companies or individual clients very often, I think 90% of the time, do not speak Norwegian. So if here is what we know and share, it is also very important to them. We send the newsletter once a month. We publish articles on Facebook and on our website. [Company J]
	"We have to exchange this knowledge with each other, because each action in a different area must be individualized due to these authorities [offices], so if we do not draw conclusions from every experience regularly and we do not all have this awareness, then simply in a particular job no one is able to complete the tasks. If they do it with one template, it will not work, so contact with the administration forces us to react here at the company level". [Company K]
Types of knowledge in the flow of knowledge	"Technological knowledge flows, knowledge related to project methodologies and knowledge derived from experience - it is always connected either with technology or with methodology and, I would say, general organizational knowledge. So all this knowledge that does not concern project management. This is, for example, financial management. In a design company, there are certain rules resulting from specific knowledge and experience". [Company C]
	"For me, procedures are such a database that anyone can consult at any time and make sure how to arrange a vacation application. These are such idiotic things that can paralyse". [Company H]

Source: Own elaboration

other members. The flow of knowledge within the company, according to the respondent, mainly constitutes conversations. The owner of the Company E was of a similar opinion, who also emphasized the importance of talks and meetings in internal knowledge flows.

In turn, the owner of the Company H pointed to two types of knowledge flows within his organization: vertical and horizontal. In his opinion, vertical flow means knowledge transfer from more experienced employees to less experienced ones. Research shows that mentoring and storytelling from personal experience are of great importance for the spread of knowledge in the organization, especially tacit knowledge (Swap et al., 2001). The second type of knowledge flow, horizontal flow, concerns employees at the same levels of the organization who should cooperate and share knowledge. For the horizontal flow of knowledge to take place, according to the respondent, communication and teamwork are necessary, which Company H is trying to implement.

The internal flows of codified knowledge were indicated by the respondent from J. In her opinion, the flow of knowledge within the organization is mainly based on e-mails, didactic materials or internal studies prepared by some employees for others. As can be seen, the flows of knowledge may refer to both explicit (codified) and tacit knowledge.

External flows can be two-way—from organization to environment and from environment to organization. In the context of the outflow of knowledge from the organization, the respondent from Company A mentioned making knowledge available to the client during the execution of the task, provided that the client is interested in receiving this knowledge. The owner of the Company G was of a similar opinion, who stated that there is no need to widely disseminate the company's knowledge, and that external knowledge flows are only related to the implemented project.

When it comes to flows from outside to the organization, the owner of Company A pointed to seminars, trainings and a broadly understood library. A similar answer was given by the owner of Company B, who, apart from training courses, also mentioned presentations or acquiring knowledge from specialists who answer the questions. The respondent from Company E broadened this spectrum with textbooks assimilated by company members.

The flow of knowledge with the external environment is important especially for companies that have to cooperate in some way with other entities (e.g. offices) or whose activities are conditioned by legal regulations (e.g. companies offering accounting and legal services). This is confirmed by the owner of Company K, who believes that the lack of obtaining external knowledge (in this case from offices) may result in a failure of the tasks.

One of the reasons why KIBS companies transfer their knowledge to the environment may be the willingness to demonstrate knowledge in a given area, which allows them to gain customers or gain their trust in the services provided by these companies. If the customer uses the knowledge of KIBS provided free of charge and it brings him some benefit, it can be expected that he will be more willing to purchase a paid service from this company.

Another division of knowledge flows was proposed by the owner of Company C, who distinguished the types of knowledge flowing in the organization. According to him, technological knowledge flows inside the organization, knowledge related to project methodologies, knowledge derived from experience (which is always connected either with technology or with methodology) and general organizational knowledge, i.e. knowledge that does not apply to project management, e.g. knowledge about financial management. According to the president of the board of Company H, an important subject of knowledge flows are various types of processes and procedures used in the organization, to which every employee should have access.

Summing up, it can be concluded that the types of knowledge flows indicated by the respondents were mainly based on the distinction between internal and external flows. Some respondents also indicated the types of knowledge transferred in the flow. There are not too many of these kinds; it can be said that the surveyed companies mainly use their internal knowledge, expanded with knowledge from training courses or professional literature. When it comes to transferring knowledge to the outside world, most companies do not see the need to disseminate their knowledge to a wide audience (the exception is Company J, which publishes magazines, newsletter and publishes materials on Facebook).

6.3.3 Knowledge Flow Between the Company and the Client

One of the most important types of knowledge flows between KIBS and the environment indicated by respondents was the flows of knowledge between them and customers. Some companies devoted a lot of attention to these flows, while others, for example, made their intensity dependent on the client's involvement or his/her willingness to acquire knowledge. On the basis of the respondents' answers, the following code categories were created related to the flow of knowledge between KIBS and the client: a competent person to contact the client, client education, regular knowledge flows to the client, making knowledge available to the client, dependence on client knowledge, matching flows to the client's needs/engagement, bidirectional flow of knowledge, place of knowledge flow between KIBS and the client and problems in the flow of knowledge with the client (Table 6.11).

In the flow of knowledge with the client, it is important to delegate an appropriate (competent) person who deals with a given task area in the company. As the respondent from N Company stated, a person who does not have the appropriate knowledge or experience will not be able to obtain from the client the knowledge necessary to provide the KIBS service and may also cause misunderstandings. The owner of Company B is of a similar opinion, who believes that it is important that the person who contacts the client is a source of information for him on how to fulfil his need.

In some cases, adequate education/providing the client with knowledge (both before and during the service) may affect the success or failure of its provision. For example, if the customer does not read the information provided by the company, he is not informed and may make a mistake. This is pointed out by the manager of Company J, who described the example of a customer who unsubscribed from the newsletter and was therefore not informed about changes in the regulations. Regular knowledge flows between KIBS and the client may also result from changes in conditions or regulations that clients must be informed about. This was the case, for example, in the case of Company A. Some companies shared a certain knowledge base with clients, and if the clients had any questions or wanted to expand this knowledge, then they reported to KIBS for help. One type of knowledge was routinely shared with the entire customer base, while another was a response to a specific customer problem and related only to a specific case.

The manager of Company F, however, used a drawing to illustrate the flow of knowledge between the company and the client (Fig. 6.1). As shown in the picture,

Table 6.11 Knowledge flows between the surveyed KIBS companies and the client

Code	Quotation
A competent person for contact with the client	"Communication with clients must be handled by experienced people who know what they want to talk about and how they want to talk. It cannot be a person who, let's say, is just learning and exploring all the secrets. And with a designer, builder or whatever, she won't know how to have a conversation and what she expects from such a person". [Company N]
	"When we talk about customer needs, it's a meeting about what needs to be done. The person who meets the client must be the source of information on how to do what the client would like. There are always some suggestions on how this can be done. The customer must be aware of what he could order. He doesn't need to know that before he comes here. He has a need and must have information on how to implement it". [Company B]
Customer education	"For example, we add someone to the newsletter, someone unsubscribes from this newsletter, and then we see that after some time this person returns and comes back, let's say, with a question that would be answered very quickly, even by reading the newsletter. So I generally think that people are doing their own thing, colloquially speaking". [Company J]
Regular knowledge flows to the client	"We communicate to clients collectively, if there are any changes, then we prepare some information e-mails and send out that from this and that day a new obligation or change of obligation is introduced, and we pass it on. And some clients, if they do not understand something, then communicate with us, we have to explain it to them more and tell them whether they are to do something or not, whether we will do something for them, *etc.* This is most often the case. However, in other cases it is a specific case, a specific case or from some area of time". [Company A]
Making knowledge available to the client	"Sometimes we organize such meetings, organize trainings, if there are such needs on the part of clients. We also provide online documentation on an ongoing basis, *i.e.* we have our own tool in the form of (...) a page available only to logged in, authorized users. There is this knowledge on different levels: from the usual end user documentation, *i.e.* a simple user manual, through the configuration manual, to advanced descriptions strictly for programmers, because there are different levels of access, and we share this knowledge with customers in this way". [Company D]
	"We have to provide the customer with the documentation, and he submits his comments. (...) When we talk to clients, they are usually oriented and then we exchange knowledge with them, we convince them that such a solution is better". [Company G]
Dependency on customer knowledge	"We have a significant number of clients for whom this topic *[subject of the service]* is completely unfamiliar and this

(continued)

Table 6.11 (continued)

Code	Quotation
	subject needs to be somehow illuminated, but it cannot be done. These are people professionally unrelated to the program, and if the clients are in any way professionally prepared in any way infrastructure companies or lawyers, then yes, we play together in one team and we work out solutions together. We know one part of the problem, they know the other. Depending on what we come to together, our behaviour may be different. But it depends on the client, because some people are not able to explain or give a general outline and give analogies in life to let them understand what we are doing at all and why such problems arise". [Company K]
Matching flows to customer needs/engagement	"Every case we do is of course to the extent we do. The client, if interested, gets full knowledge, including the fact that sometimes simple things have to be explained painfully, and we convey this, because we know that if he understands it in his own way and is convinced, our relations also change trust is changing, and that is extremely important in this job". [Company A]
	"For sure we can do a lot more *[in the area of knowledge flows]* because we don't run any kind of webinars or stuff for clients, so we don't share everything we know. The flow of knowledge is controlled manually between us and our clients. So there are meetings and there are many such meetings, we travel a lot around the world. We go to clients a lot and we are running out of time for it slowly, so I think that next year we will start thinking about how to do it to centralize it a bit and that this knowledge flows from one person and goes to wide range, not just one-on-one. At the moment, these meetings are one person from us and, let's say, five from different other companies, and this way we can boast that we know something, and it would be appropriate to brag a little more so that it would simply reach more people. We have certainly started to prepare ourselves for public speaking this year and I am sure my colleagues will start to give some lectures at all kinds of conferences, seminars and similar places". [Company H]
Bidirectional flow of knowledge	"On our part, there must be *[flow]* on how we manage the project, what the client expects, what to apply to, how to formulate his requirements towards us. This is a very important field of knowledge - requirements management. On the other hand, we get information from the client about how his business works. (...) We gain knowledge at meetings and reading documents at this client's premises". [Company C]
Place of knowledge flow between KIBS and the client	"Meetings are held mainly in our company. We rarely meet at the client's premises, if only due to the fact that during such consultations, the client is given didactic materials, which are quite a lot, and also during such consultations we can really determine what kind of entity the company will

(continued)

Table 6.11 (continued)

Code	Quotation
	be. I, going to the client, would have to take everything. And if I have a client here, we talk and during the conversation it turns out that he will be, for example, X, not Y, then I will print him all the materials concerning X". [Company J]
Problems in the flow of knowledge with the client	"It feels like you are dealing with such a person [resistant to knowledge], so I don't even try to argue then, because I know that it won't help me at all, that I will most likely lose a client and the client will stay anyway as it is, I will not change and not raise it. (...) You can't do everything at any cost either. It happened to me two or three times that I refused to provide a service because I saw that it would be hard, that everything would be wrong, the client will know better at every step". [Company M]
	"Not all clients are willing to share knowledge. We have a very open relationship with customers and I would even say that we try to educate our customers and that is part of our job. An uneducated client will never be able to appreciate what we give and what added value we bring apart from the fact that we make drawings, because we try to move away from the image of a design company that makes drawings, but rather we share knowledge and show how to avoid some kind of problems later". [Company H]
	"In 85% of the cases, he [the client], reading this description, states that there is too little written for the assignment, and he expected more, but did not disclose it at all. Well, that's the problem of how to get that knowledge from the client in such a good way. If the client turns a blind eye, he will always be dissatisfied. It happens, clients tell us that we don't have time to talk to you now, so do it, you already have some knowledge, *etc.* And it always happens that when he sees that something has happened without his consultation, then he won't like it". [Company C]

Source: Own elaboration

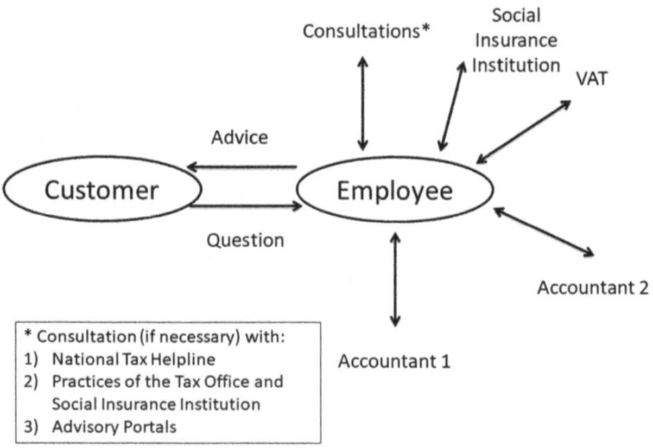

Fig. 6.1 Knowledge flows between Company F and the client (Source: Author's own illustration, based on a drawing made during the interview by the manager of the Company F)

the customer comes to the company with a problem. The company's employees consult together, e.g. Accountant 1 and Accountant 2, because one of them specializes, e.g. in full accounting, and the other deals with income and expense books. The team also includes a person who knows VAT well, and an employee who handles Social Insurance Institution, human resources and payroll. Communication takes place in both directions, with employees, the facts are communicated to them and they influence the shape of advice for the client and his decisions. The circle of people involved in making decisions and advising the client is very large. There is a person who integrates knowledge from various sources and is responsible for the overall preparation of the advice. The advice is determined inside the company and then passed on to the client. If any information is missing, the client is contacted to collect all the necessary data before advice is given.

The flow of knowledge is often necessary at the beginning of the service provision or when determining its subject/scope. The further flow of knowledge between KIBS and the customer may depend, inter alia, on the level of knowledge and on the ability of the client to absorb new knowledge. This dependence is indicated by the owner of the Company K, who says that some customers are not able to understand the essence of the service and only report the problem, hoping that it will be resolved by KIBS.

Interestingly, the intensity and the way in which knowledge flows between KIBS and the client are organized seem to depend in part on the size of KIBS. Company H, which is close to a medium-sized company (40 employees), begins to consider transferring knowledge to its customers in a more structured way, allowing it to reach a wider group of recipients. As the respondent from that company stated, at present the knowledge flows to customers in a one-on-one or one employee basis from KIBS towards several employees from different customer companies, but this solution seems to be insufficient. As stated by the President of the Management Board of Company H, the current formula of transferring knowledge to clients, based on meetings, begins to be difficult to maintain and it is necessary to introduce other solutions, e.g. webinars, conferences, seminars, etc.

The flow of knowledge will also depend on the type of KIBS service provided. For example, the Company J offers its services most often in the form of consultations, which can be either in writing or in the form of a meeting. Although there is information in the literature that it is KIBS that sends its employees to the client company (Bettiol et al., 2015), in the case of Company J it is the other way round—consultation takes place at the headquarters of KIBS. This is due to the need to provide the customer with materials that depend on the customer's situation or the type of problem. In general, where knowledge flows between KIBS and the customer may depend on the following factors:

- *number of people from individual organizations* (KIBS companies and the client's companies) participating in the meeting: e.g. if the meeting is attended by more people from the client's side, it will be more convenient to meet at his office;

- *space conditions:* e.g. whether KIBS can accommodate a given number of employees from the client company;
- *the distance between KIBS and the client:* e.g. if it is very large, the flow can take place with the support of technologies such as Skype or instant messaging;
- *the importance of the environment in which the meeting takes place:* for example, if KIBS needs to observe or consult different people from the client's firm, an appointment with the client's firm may be necessary;
- *access to materials necessary to provide the service* (e.g. reports, balance sheet, customer lists, etc.): is it easier at KIBS or at the client company.

Knowledge flows with customers can take various forms and run through various channels. Some companies try to use a variety of channels to transfer their knowledge to clients. For example, Company J uses Facebook to answer questions from customers and also provides basic advice by e-mail and phone. Some standard solutions are offered free of charge, and solving complex customer problems that require individual approach is paid.

The owner of Company C pointed out the necessity of bidirectional flows. According to him, the flow must be in two directions: from the client to the KIBS company and from the KIBS company to the client. KIBS should obtain knowledge from the client about how his company operates. Thanks to this, it is possible to propose a better solution, more closely corresponding to the client's requirements and the conditions in which it operates.

Knowledge flows between KIBS and customers can be difficult when customers feel they know best and are resistant to knowledge. Such a situation was mentioned by the owner of Company M. In her opinion, it happens that clients do not want to accept knowledge from KIBS because they think that they know everything best themselves. In such a situation, persuading the customer does not make sense, and the service may not be possible. Knowledge flows from KIBS to the customer and vice versa are necessary for the service to be prepared and delivered in the right way. Lack of willingness to absorb knowledge on the part of the client may prevent this. The owner of Company H also pointed out the problems with the flow of knowledge with the client, according to whom the client's failure to acquire knowledge from KIBS may result in the inability to appreciate the importance of the service offered. In turn, the owner of Company C points out the need to transfer the client's order to a business task. In his opinion, the formulation of such a task consists in determining what can be done in the prepared system and what not, how it will work, etc. After preparing a description of such a task, the client should accept the description, stating that it is consistent with what the client received. This is not always the case, because sometimes the client does not disclose his knowledge or does not pay attention to knowledge flows and then problems arise. In the absence of flow, there is a risk of misunderstanding and, consequently, customer dissatisfaction with the service.

It is important to remember about the conversion of knowledge during/before/after the flow of knowledge—knowledge changes from overt to hidden, from hidden to overt, from overt to overt, from hidden to hidden. In addition, knowledge is converted into information and data and the other way around—sometimes it is necessary to simplify knowledge, reduce it to the form of data, so that the client can understand a given issue or take appropriate action. An example is Company J, which has a very good knowledge of the legal regulations in country X[4]—does not transfer this knowledge to the client, but analyses his problem and advises the use of solution A or B, explaining the differences between them. The client will not receive all the knowledge, but will receive information on what exactly he should do, e.g. register the company in a specific form. This is specific information—you should do it anyway because it is the best solution for your situation.

To sum up, it can be stated that the knowledge flows between KIBS and the client may be of different intensity and depend on many factors:

- *there are companies that focus on the flow of knowledge with the client in order to be able to function more easily—they make the client aware of them, show them various options, it is easier for them to talk to the client;*
- *there are companies that do not transfer knowledge to the client, creating a service on their own based on their knowledge and competence, and expect the client only to articulate the problem;*
- *there are companies that constantly have to exchange knowledge with the client, otherwise the service will not be properly provided and the client will be dissatisfied;*
- *the lack of the ability/willingness on the part of the client to participate in the flow of knowledge means that KIBS may refuse to execute the order/resign from the client.*

This part of the work discusses the perceptions of knowledge flows by respondents from KIBS companies, as well as the types of knowledge flows that occur in their organizations. Particular attention was paid to the knowledge flows between KIBS and the client, illustrating the respondents' different approaches to these flows and the factors they may depend on. The further part of the monograph presents the relationship between knowledge flows and workflows, the use of technology and various tools for knowledge flows, as well as the determinants of knowledge flows within the organization and between the organization and the environment.

6.3.4 Knowledge Flows and Workflows

Another aspect examined was the relationship between knowledge flows and workflows. The respondents were asked if the flow of knowledge is related to the workflow, and if so, how. Some companies indicated that the flow of knowledge is closely related to the flow of work. For example, in Company A, the flow of knowledge cannot be separated from what the employees do. The situation is similar

[4]The name of the country was not provided due to the possibility of identifying the company.

in Company M, where without the appropriate flow of knowledge it is impossible to carry out tasks at the required level. The owner of Company F also stated that the flow of knowledge is continuous because there are new problems to be solved or new issues to be discussed all the time. Thus, it can be concluded that the flow of knowledge is in what sense necessary to carry out tasks in KIBS companies. Detailed opinions of selected respondents are presented in Table 6.12.

Another question that was attempted was whether the knowledge flow in the surveyed companies is sequential (that is, after or before the workflow) or parallel (that is, during the workflow). We deal with sequential flow in Company I, where the flow of knowledge is closely related to the assigned work/task—it occurs before the implementation of a given task, so that the employee knows what and how to do it. In some companies, the flow of knowledge follows tasks that are to be performed by different people/departments. For example, in Company D, the flow of knowledge is transferred along with the various stages of service implementation. First comes the workflow, then the knowledge flow, then the workflow again, and then the knowledge flow, etc. A similar situation takes place at Company K, where the workflow must be preceded by the knowledge flow. Also in Company C, the flow of specific knowledge (related to a given project) must take place when receiving the task in order to be able to perform it. In turn, in Company H, the flow of knowledge takes place both at the beginning of the project implementation, when anyone can ask any question, and at the end of the project implementation, so that the conclusions of this implementation can be analysed.

In some companies, the flow of knowledge was parallel to the flow of work. The constant flow of knowledge during work was indicated by the representatives of Companies E and Companies J, in which employees constantly contact each other, transferring knowledge either directly (if they are located close to each other) or via communicators. In Company M, too, the workflow and the flow of knowledge occur simultaneously, because it is necessary to constantly consult on the implementation of individual tasks. Also, according to the owner of Company B, knowledge flows must occur in parallel with workflows, because each project is new and requires ongoing problem solving, which is related to the search and flow of knowledge.

In turn, in Company L, the flow of knowledge resulting from hierarchical relationships was indicated. For example, if an employee returns from training and his supervisor decides that he has to share this knowledge with other employees, he instructs the employee to train his colleagues. Company L belongs to medium-sized companies, so it can be expected that the knowledge flows are more structured/hierarchical in it. Only one company (Company N) indicated no relationship between knowledge flow and workflow.

In addition to the relationship between the flow of knowledge and the flow of work, the president of the board of Company H also pointed to the close relationship between the flow of knowledge and the flow of resources. In his opinion, the fact that resources in his company are assigned to different projects and tasks allows knowledge to flow between employees, because the knowledge gained during the implementation of one project is transferred to other employees during the implementation of another. The perception of resources as a pool also allows employees to be made

Table 6.12 Knowledge flow versus workflow and resource flow in the opinion of respondents from KIBS companies

Code	Quotation
Close relationship between knowledge flow and workflow	"Here, in principle, this flow of knowledge, detached from what [employees] do, is practically non-existent. He is strictly, strictly connected with what they do". [Company A]
	"He is very bound, because if this knowledge is not at the appropriate level, the implementation of tasks is not at the level it should be". [Company M]
Sequential flow of knowledge	"Special knowledge goes along with the task that a programmer or analyst or tester gets, because, *e.g.* a tester also needs to know what to test, so somehow he has to get this knowledge about this episode or area of software". [Company C]
	"[knowledge flow] goes from programmers to implementers to our implementation partners. Naturally, this knowledge flow is related to the workflow. Because they first have to create something so that it can be implemented, and this must also be followed by specific knowledge, not only the software itself, but also documentation, as well as technical manuals, service manuals. How to do it, how to configure it, how to connect it with some external systems". [Company D]
	"At the beginning of the project, we have a kick-off meeting. There are no stupid questions during the kick-off meeting (...), so you can ask everyone about everything, even if the question is to be completely idiotic, it may turn out that there is something in it. Then, after moving to the first cycle of the project, we meet again and close the project. We usually do this meeting in an informal way, that is, we go downstairs for a beer and sit down and talk about what happened, why it happened, what conclusions to draw from it for the future, and that stays in this group". [Company H]
	"When a task is assigned, then the knowledge of how [task] is to be done is immediately imparted. So there is immediately flow". [Company I]
	"It is impossible to hand over the work without transferring knowledge, otherwise it is completely wandering in the dark. (. . .) The condition for handing over the work is the transfer of knowledge". [Company K]
Parallel flow of knowledge	"It happens on a regular basis as each project is new. Very often there are new elements and you have to look for solutions on how to do it, so in many projects it is the case that performing a given task is related to the search for knowledge and the flow of this knowledge". [Company B]
	"It is necessary to update this work all the time. Even if there is only one person responsible for the project, he consults all the time, because it is a work that consists of some stages, he has to consult these stages anyway, but he

(continued)

Table 6.12 (continued)

Code	Quotation
	also has a lot of scope for such independent action". [Company M]
The flow of knowledge and the flow of resources	"The *[company]* structure is such a pool of resources, and these resources flow, and then it turns out that the person who learned the knowledge on a given project passes that knowledge on to another project. I would say that the flow of knowledge ties in very nicely with this flow of resources. We had studios earlier in the organizational structure. Such a big change we did, and these labs were responsible for part of the job, which is say we have a bottle and one lab was responsible for creating the cap, and the other lab was responsible for creating the bottle. Then it turned out that we had the greatest bottle, the greatest cap, only the bottle does not fit the cap. Therefore, we decided that the workshops should be closed, destroyed and never returned to them again, because they were such funny conflicts. The cap doesn't fit the bottle, but we have a great bottle. Well, our cap is also great, but it does not fit, and that's why these flowing resources have kind of taught people to take responsibility for everything. So look wider and not focus on these retail things, because these retail things can always be improved, and the big ones are difficult, when the project is over, go back to the beginning and improve, start all over again". [Company H]

Source: Own elaboration

aware that they are part of the whole and must take responsibility not only for their part of the work, but also for the implementation of the entire order.

To sum up, it can be stated that in the surveyed KIBS companies, knowledge flows are closely related to workflows. Some companies view knowledge flows as sequential, before or after workflows, while others see them as parallel, taking place at the same time as workflows. In the next part of the study, the issues related to stuck knowledge and the ways of counteracting this situation are discussed.

6.3.5 Knowledge Stuck and Ways to Counteract This Phenomenon in KIBS Companies

Knowledge stuck can be defined as a lack of access to knowledge in an organization—although someone has it, it is not known who it is and how it can be accessed. Some respondents stated that in their organizations the problem of knowledge stuck does not exist (Company I, Company G, Company K, Company J, Company B). This is due to the fact that getting stuck could have consequences for the entire team/ company. This situation was indicated by the owner of Company A, according to whom all employees are vigilant and appreciate the importance of some kind of knowledge, which may turn out to be very important. Therefore, there is no situation where something "gets stuck" and is potentially needed by someone to work. An important element here is shared responsibility for work and tasks in the company. The owner of Company D is of a similar opinion, who was not able to recall that someday knowledge stuck somewhere in the organization. He only pointed to a situation where the solutions implemented some time ago are difficult to modify, if the person responsible for them no longer works in the company and has not left the relevant documentation, but such problems are not directly related to knowledge stuck in the organization itself, but to its loss with the departing employee. Selected quotes from respondents illustrating the problem of knowledge stuck and ways to counteract this phenomenon are included in Table 6.13. However, these kinds of problems are not directly related to knowledge being stuck in the organization itself, but to its loss with the departing employee.

In the remaining companies, the problem of knowledge stuck appeared in various ways and was caused by different factors, e.g. lack of knowledge flow between employees, inappropriate addressing of knowledge (to the wrong person), reluctance of an employee to share knowledge and keep knowledge for himself (the so-called knowledge hoarding). and knowledge hiding) or problems with the client. An example of stuck knowledge was shared, for example, by the president of the management board of Company H—the organization was looking for a given knowledge, and not one person who had this knowledge was asked. A similar situation took place in Company C, where a solution to a problem that turned out to be previously solved in another project or task was also sought. The respondent from Company C pointed to the dilemma related to this kind of stuck knowledge—to what extent the knowledge in the company should be formalized and to what extent. On the one hand, access to codified knowledge often allows for faster access to the sought knowledge, and on the other hand, codification is time-consuming and costly. In this case, knowledge sharing is crucial as it helps organizations promote best practice and reduce unnecessary efforts related to re-learning or re-inventing the wheel (Lu et al., 2006).

In connection with the problem of knowledge being stuck, the respondents suggested the following solutions:

- searching for an addressee of knowledge (in the event that the knowledge reaches the wrong person in the organization);
- making employees aware of why knowledge should be shared and how it affects the organization's performance;
- creating an appropriate climate and culture conducive to sharing knowledge (e.g. by spending free time together by employees, where they can integrate and learn more about each other);
- visual methods of spreading knowledge;
- a competency matrix and a data bank, allowing you to quickly find a person who may have the missing knowledge;
- organizing short, daily meetings (standing) so that each employee can tell what problem they are struggling with and possibly find someone who solved it earlier.

Table 6.13 Knowledge stuck—occurrence and prevention in the surveyed KIBS companies

Code	Quotation
No knowledge stuck problem	"The only way to get stuck is that there was no interest or there was a problem with only one company, one employee. On the other hand, if there is something generally important on such a forum for us, it is always this flow". [Company A]
	"I don't remember such a situation. There are such situations, *e.g.* there is a system implemented by a client that someone implemented some years ago, made some specific solutions in this system, in this configuration, and he did not document it, he did not share it anywhere. And the client after some time or that partner took up another thing, left, for example, or changed his profession or profession altogether. There were such cases and you had to really somehow find out about it yourself, in what was done there, and it just takes more work to analyse it, if there is no special documentation prepared for it or a person who is willing to share this knowledge". [Company D]
Knowledge stuck	"Yes of course. We have had situations at home more than once when nobody knows, and it turns out that the boy knows two desks away, but somehow someone forgot to ask him". [Company H]
	"One of the problems in projects, especially in large ones, was that someone trìed to solve a problem that could not be solved, a programmer, But it turned out that in the same room or in the next room there was someone who had already solved the problem or in another project, or just before, and knows how to do it right away, and this is a discussion of how much knowledge to formalize and how much not". [Company C]
	"It happens that someone wants to keep knowledge for himself and then will not share it". [Company M]
Ways to counteract	"More and more we go into the visual methods of spreading knowledge and we start, you can see here, the pages are hanging somewhere and so on. Arrays appear. We're going to visualize just the flow of knowledge. Who knows who might know, the competence matrix also answers this question in some way. I am looking for knowledge, I open to myself, I look at who can help me. I think that here visualization [is the solution], that is, to create a visual data bank in which you will know what you can ask whom to ask. In my opinion, such informal communication plays a very important role, so it is good to provoke employees to spend their free time together from time to time. Then it turns out that someone has a strange hobby and this hobby can be used for something. We have a friend who paints figurines and makes money for herself and is good at it. Then it turned out when we started to create 3D prints of various elements for marine gyms, she painted them beautifully. So that there were moments in which life outside the company turned out to be cool elements that can be brought to the company". [Company H]
	"This applies especially to such detailed cases, often a programmer encounters an exceptional situation in programming, when solving some problems. It was not worth describing such a situation because it is truly unique. And he can spend days or even weeks doing it, and later find out that someone in the next room is sitting and had this knowledge. The solution to avoid this is to have short meetings every day. Necessarily standing, where the whole team meets and talks about a problem they

(continued)

Table 6.13 (continued)

Code	Quotation
	had, and at that moment someone can say: 'but I solved it a long time ago', this is a way that makes it very easy. At some point, routine appears again and people do not like going to such meetings anymore, because nothing spectacular happened, which again needs patience and knowledge, how to cultivate such meetings and make people willingly participate in them. That is why it is standing so that people can see that they are not sitting down, so that people are convinced that this meeting will be short". [Company C]

Source: Own elaboration

6.3.6 Technologies and Tools in Knowledge Flows

Another aspect examined was the use of various types of technologies and tools for knowledge flows. When it comes to knowledge flows between KIBS and the client, they often take place in face-to-face meetings. The importance of a direct relationship with the customer as a characteristic of KIBS companies has been emphasized in various studies (Czarnitzki & Spielkamp, 2003; den Hertog, 2000; Muller & Zenker, 2001). Interestingly, as shown in Table 6.14, most of the surveyed companies showed high homogeneity of the technologies used for the flow of knowledge. Almost all companies indicated e-mails, telephones, instant messaging and face-to-face conversations/meetings. This is partly in line with the results of the research conducted among SMEs. According to Centobelli et al. (2018), e-mail is the most frequently used knowledge management tool in such companies. Moreover, the authors of this study postulate that there is a "digitization border" that companies from the SME sector do not want to cross, because they do not see such a need. Knowledge in these companies is often hidden in the minds of employees and its digitization would take too much time and require significant financial outlays.

Some companies emphasized the importance of conversations as the only "technology" used for knowledge flows (e.g. Companies A and K). Some companies also indicated various types of solutions, such as Skype, intranet, Jira,[5] HipChat or Yammer. While the question was about technology, most companies emphasized the importance of traditional conversations. For example, the president of the board of Company H stated that apart from the tools used, meetings with employees are very important.

> "We meet our employees regularly, I think this is the most important thing. Honestly, it is probably time to go back a little bit, which in our opinion means

(continued)

[5]Jira is Atlassian's proprietary bug tracking and project management software. The name is the abbreviated form of the word Gojira, which is the Japanese name for Godzilla. Jija is used in many projects, such as Wikipedia (source: https://pl.wikipedia.org/wiki/JIRA).

Table 6.14 Technologies and tools in knowledge flows used by the surveyed KIBS companies

Code	Company
Conversations	Company A, Company B, Company E, Company F, Company G, Company H, Company I, Company K, Company N
E-mails	Company A, Company B, Company E, Company F, Company G, Company H, Company I, Company J, Company L, Company M, Company N
Telephones	Company G, Company J, Company N
Internet	Company I
Skype, instant messaging	Company H, Company J, Company M
Intranet	Company L
Jira	Company E
HipChat	Company E
Yammer	Company H

Source: Own elaboration

that modern media are becoming less and less important. There are just too many of them and people are dying a little in them. And more and more important are meetings and going out for a beer and simply direct communication". [Company H].

This is an interesting voice in the discussion, especially as the development of technology may entail risks. On the one hand, we observe a growing interest in big data and the Internet of Things, as well as in the development of artificial intelligence and on the other hand, as you can read in Technology doesn't make us' Appy, technology is a double-edged sword. It quickly gave us unprecedented access to ideas and information and connected us with people all over the world, but it has a price and it is young people who are most affected by the potential negative effects of technology.[6] Ones need to keep in mind that they will be the workforce in the future and encouraging them to use traditional means of communication and return to their roots can be a healthy alternative to the excessive use of technology.

However, it should be borne in mind that despite the potential threats, the use of various technologies (e.g. e-mails) for the flow of knowledge brings benefits and is often dictated by the needs of the surveyed companies. For example, the owner of Company A points out that the use of e-mail is useful if the knowledge flow content needs to be preserved/archived in some way. The owner of Company F is of a similar opinion, as she documents the flow of knowledge with the client which are key and potentially problematic.

[6]Whitehead K. (2014), Technology doesn't make us' Appy: young Hongkongers "depressed by social media", http://www.scmp.com/lifestyle/family-education/article/1653176/how-social-media-having-detrimental-impact-young-people (access date: 20/03/2018).

The use of technology also makes it easier for KIBS companies to overcome the distance between them and their customers—they do not necessarily have to meet the customer face to face, they can use, for example, Skype. Technologies also enable communication and work in teams (such as tools such as HipChat or Yammer) and project management (Jira). It is worth noting that in the case of the surveyed companies, the use of these technologies resulted from real needs (e.g. clients located too far away to be met face to face, several offices dispersed across Europe, the need to document contacts with the client, etc.).

6.3.7 Factors Influencing the Flow of Knowledge

The respondents were also asked to indicate factors influencing, in their opinion, the flow of knowledge. When it comes to flows between KIBS and the customer, certain conditions must exist for this type of flow to be effective. These conditions include: an appropriate level of generality of the knowledge transferred, the specification of knowledge, articulation of needs by the client, elimination of problems with the recipient of knowledge. Table 6.15 presents quotes from selected respondents' responses regarding these factors.

According to the owner of Company C, it is important to present the client with specific knowledge at the appropriate level of generality. If the customer receives too much knowledge or it is too detailed, the creation of the service may be difficult. At the same time, the lack of a reliable conversation with the client and the lack of explanation of the matter at the appropriate level of detail may result in problems with the service. It is important to properly articulate the knowledge and make sure that the recipient of the knowledge appropriately assimilates it. This type of case was pointed out by the manager of Company J. It can therefore be concluded that KIBS companies must balance between providing the client with too much knowledge (or too detailed knowledge) and not informing him or her about the subject (details) of the service.

At the same time, the owner of Company M pointed out to the problems in articulating the needs of the clients. In her opinion, clients often do not know what they expect from KIBS, they have some general understanding, but are unable to define the scope of the service in detail. KIBS must then organize meetings a number of times, show examples of solutions and thus come to the actual needs of the client. In turn, the manager of Company J pointed to a problem in the flow of knowledge related to the recipient of knowledge. In some cases, the recipient of knowledge is not ready or willing to assimilate the knowledge. According to the respondent, the transfer of knowledge is sometimes ignored and the recipient does not want to absorb it. The respondent gave an example when customers unsubscribe from the newsletter prepared and sent out by Company J.

Conversely, when it comes to the factors influencing the knowledge flow within the organization, they can be divided into two categories:

Table 6.15 Factors influencing, according to the respondents, the knowledge flow between KIBS and the client

Code	Quotation
Appropriate level of generality of the transferred knowledge	"It's like that when we talk to the client, we cannot go down to the level of absolute details during every conversation, because we would never end this conversation, and we can talk at a high level of abstraction, because it turns out that we do not solve specific cases., and again we have to find the right level, and sometimes it's just like the whole project can't find it, so it raises various problems". [Company C]
Refinement of knowledge	"Sometimes some understatement *[affects the flow of knowledge]*, because on the part of clients you can observe that the client will say something, say 50% of what he should say, and the other 50% come out later and what now? "Why didn't you say this before?" "Oh, because you didn't ask me". And I say: 'if you knew that certain issues might happen in the future, or they are already partially happening, then we had to tell us, then we would have handled it a little differently. So I think sometimes such errors, maybe not errors, but inaccuracies in communication, make it difficult. Sometimes the client fails to specify what exactly he needs. Then there is also a blind shooting, reading tea leaves". [Company J]
Articulating needs by the client	"Ordering persons often have a problem with conveying what [service] should look like, what it should be, so you have to hold a number of meetings, talk and show on examples. Someone shows what they like, what solutions, what functionalities, examples of subpages". [Company M]
Elimination of problems with the recipient of knowledge	"Sometimes people make it difficult for themselves. In the sense of the person to whom we want to direct this knowledge. Because, for example, we add someone to the newsletter, someone unsubscribes from this newsletter, and then we see that after some time this person returns and comes back, let's say, with a question to which they would find an answer very quickly, even reading the newsletter. So I generally think that people are colloquially speaking, 'they are doing it uphill". [Company J]

Source: Own elaboration

- resource/human: concerning people participating in the flows;
- technical: related to the technical aspects of knowledge flows.

The first human factor influencing the flow of knowledge and indicated by the respondent was trust. As the owner of Company A stated, trusting other employees who may have important knowledge to pass on is very important. Some companies (e.g. Company D) pointed to the problem with human resources, which are not enough and the time they have. Employees are sometimes so overloaded with their daily work that they are unable to participate in the flow of knowledge on an ongoing basis, because they lack time for it. The solution to this problem could be increasing employment, but many small businesses cannot afford it. One of the characteristics of the SME sector is limited access to human and financial resources (Torrès & Julien, 2005). Details on the factors influencing the flow of knowledge within KIBS companies are presented in Table 6.16.

An important human factor is the atmosphere and characteristics of people participating in the flow of knowledge (Company K). The president of the management board of Company H also pointed to the role of atmosphere, relations and communication in the company. The place for integrating employees is a shared kitchen, where they can not only prepare meals, but also participate in the flow of knowledge. Another factor is the willingness to cooperate and the desire to expand knowledge. This factor was pointed out by the owner of Company E and the manager of Company F. Employees must, above all, be willing to cooperate with each other and with their superiors and must not be afraid of new challenges and broadening their knowledge. Moreover, according to the respondents from Companies B and F, it is also important to properly motivate, often non-financial. Both respondents indicated non-material incentives, such as, for example, integrating events (e.g. rafting).

When it comes to the technical factors influencing knowledge flows, most companies did not perceive them as problematic. Some companies said explicitly that the technical limitations did not apply to them because they had different tools (e.g. Company D). Knowledge flow channels play an important role and ensure that knowledge reaches all those who need it (Company L). The availability of knowledge flow channels is one of the elements determining the flows, also according to the respondent from Company F, who pointed to the need to access professional Internet portals from which employees can obtain the necessary knowledge.

Other technical factors influencing the flow of knowledge in KIBS companies included, among others, organization of an open space office.[7] This type of office organization was indicated by the respondents from Company F and Company K. This solution has its advantages and disadvantages. The advantages include easier communication between employees and team integration. The disadvantage of this type of solution is its inadequacy for people who cannot work or focus in such conditions—their effectiveness may suffer.

When it comes to factors influencing the flow of knowledge between KIBS companies and the environment, one company indicated financial resources that hinder or support the flow of knowledge. According to the owner of Company K, the lack of financial resources makes it impossible to participate in trainings and gain knowledge from outside the organization. Among the factors supporting the flow of knowledge with the environment, the respondent from Company B included contact with the market, thanks to which it is possible to obtain knowledge from various

[7] An open space office is a large-space office in which employees' desks are arranged in one large space, without being divided into separate rooms or offices.

Table 6.16 Factors influencing, according to the respondents, the flow of knowledge within KIBS companies

Code	Quotation
Resource factors	
Trust	"Trusting certain people who can say something smart when everyone is wrong. Sometimes I am also wrong". [Company A]
Limited resources and time	"They make every day work difficult and there is always not enough time as you would like". [Company B]
	"Resources for sure. This means that because we are a small company, there are times when there are simply more orders and such current work, and this is the main difficulty. It is not always enough time at a given moment to exhaustively keep this documentation somewhere. So that's really the main problem (. . .). The main limitation is resources - the number of people employed and the time they have at their disposal". [Company D]
The atmosphere and characteristics of employees	"The atmosphere here definitely influences the characteristics of individual employees". [Company K]
	"We have a very relaxed atmosphere, it is so very homely. We have a kitchen, the kitchen is the best party as always, so we just try not to create artificial divisions. Everyone admits they don't know something, including management. Everyone is open to communication. Virtually all doors are open. Anyone can talk to anyone. We do not create artificial barriers on the basis of 'Good morning, Mr. President, I would like to', but rather, we focus on free communication between everyone. We organize daily meetings. There are briefings. We have stopped looking at projects in a cascade. We started to look at it as a whole and this is probably the most important change when it comes to communication. There were also big problems because people didn't want to talk. People wanted to get their job done, hide behind a desk and get their job done".
Willingness to cooperate and the desire to expand knowledge	"The employee's willingness to expand their knowledge. I think that with us, everyone is so eager to work, they like to do it". [Company E]
	"Willingness, willingness to cooperate. You have to educate employees on this topic all the time. The willingness to cooperate to show that now we have such a turnover, we had a much smaller turnover, but we are not afraid, the management is not afraid to enter new topics, because the team is competent and wise". [Company F]
Adequate motivation	"The incentives are unfortunately not economic, financial, but friendly. Through integration. We do fun together, celebrate a birthday together, canoeing or some mushroom picking. Such integration events, so that many employees want to work with us. I keep in touch with 90% of those who have passed away, they come to our meetings". [Company F]

(continued)

Table 6.16 (continued)

Code	Quotation
Technical factors	
Technical limitations	"We don't have any technical limitations right now because we use different tools to share it". [Company D]
	"Availability of professional internet portals". [Company F]
	"When it comes to inside the organization (...) e-mails are certainly the basic thing and the assumption that an e-mail that should reach everyone is sent to everyone in advance (...)". [Company L]
Organization of the office	"Organization of the office and constant exchange of views on every topic, all the time, all the time. The office is structured in such a way that it has five rooms, but all doors are removed, we have removed the door at all. I am giving advice in an open transition room and the employees are telling me because there is such a huge amount of knowledge". [Company F]
	"There are three or four people sitting there, and they have to [exchange knowledge] with each other. It is a friend who knows is at hand, so he shares this knowledge on a regular basis. This space somehow helps that we are all close at hand". [Company K]

Source: Own elaboration

sources, including from producers of various products, as well as competitors, or during conferences and trainings.

As for the ways of improving the flow of knowledge, they included, among others an unrealistic extension of the day (so that employees can complete all tasks), as well as increasing employment. In turn, the manager of Company J proposed the following ways to improve the flow of knowledge, which she successfully uses at home:

- weekly recurring meetings;
- greater trust of managers towards subordinates: it is important to trust the employee and entrust him with duties (the so-called empowerment) so that he can perform them independently, and if he encounters a problem, he will ask a more experienced person for help/advice;
- elimination of the so-called the rat race, which makes employees reluctant to share knowledge: any person who acquires this knowledge wants to have it only for himself, because he wants to "shine";
- Spreading awareness that all employees play to the same goal, which makes them more willing to share everything they learn with other employees.

6.4 Summary

Summing up, it can be stated that the main factors influencing the flow of knowledge were related to the employees of the surveyed KIBS companies, perceived as knowledge capital. This is in line with the picture obtained in the previous section on technologies and tools in knowledge flows—most respondents pointed to the importance of conversations and direct exchange of knowledge between employees, and technologies only supported some flows and were selected depending on the needs of a given company. Therefore, taking into account the importance of conversations and direct contact between employees, the respondents very often indicated those concerning employees as factors influencing the flow of knowledge. Technical factors were not a problem, the role of technology and tools in knowledge flows is relatively small, which can be quite surprising.

This chapter presented the characteristics of the surveyed companies, followed by a cross-sectional analysis of case studies. It presented successively issues related to knowledge, knowledge management and then knowledge flows. In the next chapter of the work, the results of the research were summarized and final conclusions were offered.

References

Bettiol, M., Di Maria, E., & Grandinetti, R. (2015). Service customisation and standardisation in combinatory knowledge-intensive business services. *International Journal of Knowledge-Based Development, 6*(3), 241. https://doi.org/10.1504/IJKBD.2015.072812

Cardon, M. S., & Stevens, C. E. (2004). Managing human resources in small organizations: What do we know? *Human Resource Management Review, 14*, 295–323. https://doi.org/10.1016/j.hrmr.2004.06.001

Centobelli, P., Cerchione, R., & Esposito, E. (2018). Boundaries of digitalization – Why companies are still using e-mail and other traditional tools to manage their knowledge – And will they continue? In F. Schupp & H. Wöhner (Eds.), *Digitalisierung im Einkauf* (pp. 27–44). Springer Fachmedien Wiesbaden GmbH. https://doi.org/10.1007/978-3-658-16909-1_3

Corrocher, N., Cusmano, L., & Morrison, A. (2009). Modes of innovation in knowledge-intensive business services evidence from Lombardy. *Journal of Evolutionary Economics, 19*, 173–196. https://doi.org/10.1007/s00191-008-0128-2

Czarnitzki, D., & Spielkamp, A. (2003). Business services in Germany: Bridges for innovation. *The Service Industries Journal, 23*(2), 1–30. https://doi.org/10.1080/02642060412331300862

den Hertog, P. (2000). Knowledge-intensive business services as co-producers of innovation. *International Journal of Innovation Management, 4*(4), 491–528.

Lu, L., Leung, K., & Koch, P. T. (2006). Managerial knowledge sharing: The role of individual, interpersonal, and organizational factors. *Management and Organization Review, 2*(1), 15–41. https://doi.org/10.1111/j.1740-8784.2006.00029.x

Muller, E., & Doloreux, D. (2009). What we should know about knowledge-intensive business services. *Technology in Society, 31*(1), 64–72. https://doi.org/10.1016/j.techsoc.2008.10.001

Muller, E., & Zenker, A. (2001). Business services as actors of knowledge transformation: The role of KIBS in regional and national innovation systems. *Research Policy, 30*(9), 1501–1516. https://doi.org/10.1016/S0048-7333(01)00164-0

Panagiotakopoulos, A. (2012). Staff "poaching" in the small business context: Overcoming this key barrier to training. *Industrial and Commercial Training, 44*(6), 326–333. https://doi.org/10.1108/00197851211254752

Schindler, M., & Eppler, M. J. (2003). Harvesting project knowledge: A review of project learning methods and success factors. *International Journal of Project Management, 21*(3), 219–228. https://doi.org/10.1016/S0263-7863(02)00096-0

Swap, W., Leonard, D., Shields, M., & Abrams, L. (2001). Using mentoring and storytelling to transfer knowledge in the workplace. *Journal of Management Information Systems, 18*(1), 95–114. https://doi.org/10.1080/07421222.2001.11045668

Torrès, O., & Julien, P. A. (2005). Specificity and denaturing of small business. *International Small Business Journal, 23*(4), 355–377. https://doi.org/10.1177/0266242605054049

Zieba, M., Bolisani, E., & Scarso, E. (2016). Emergent approach to knowledge management by small companies: Multiple case-study research. *Journal of Knowledge Management, 20*(2), 292–307. https://doi.org/10.1108/JKM-07-2015-0271

Discussion of the Research Results and Conclusions

7

7.1 Research Results: Discussion

The surveyed companies offered various types of services, such as: advertising, accounting, research and development, technical design, IT services as well as geodetic and geological services. Some respondents believed that their services were mostly non-routine, while others pointed to the predominance of routine services. The position in the literature according to which KIBS companies offer only non-routine services (Corrocher et al., 2009; Muller & Doloreux, 2009) turned out to be not entirely true. Some companies also indicated offering standardized (industrialized) services that could be tailored to the individual needs of customers. Generally speaking, the types of services offered by the surveyed entities can be included in the classification presented in Fig. 7.1.

The results concerning the organizational structure of the surveyed companies turned out to be surprising. Although the natural choice for companies from the SME sector is often a flat structure (Durst & Edvardsson, 2012; Saini & Budhwar, 2008), conducive to the exchange of knowledge and flexibility (Daft, 2010, p. 25), in the surveyed organizations, sometimes a hierarchical or mixed structure was indicated. The hierarchical structure resulted, for example, from legal responsibility for decisions made or the level of knowledge possessed by employees. One company indicated a hierarchical and process structure, conditioned by the free selection of resources for the implementation of individual projects. Thus, it can be concluded that the organizational structure of the surveyed companies corresponded to their needs in terms of management and operation and, contrary to natural expectations, it was not always a flat structure.

When it comes to knowledge essential for the functioning of the surveyed KIBS companies, all companies indicated professional (specialist) knowledge, which is the basis for creating KIBS services. Many of the definitions of KIBS services presented in Chap. 2 indicated the special importance of this type of knowledge (Koch & Strotmann, 2008; Miles et al., 1995). The second type of knowledge important for

M. Zieba, *Understanding Knowledge-Intensive Business Services*, Knowledge Management and Organizational Learning 10,
https://doi.org/10.1007/978-3-030-75618-5_7

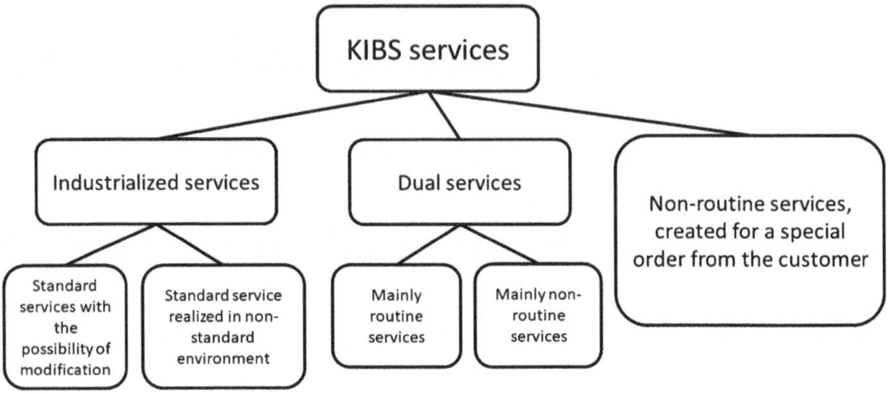

Fig. 7.1 KIBS services offered by the surveyed companies. Source: Author's own illustration

KIBS companies, indicated by the respondents, was knowledge derived from experience. It is this experience that often allows KIBS companies to better understand customer problems and offer them solutions that meet their needs.

While all the surveyed companies agreed that knowledge is their source of competitive advantage, there were different perceptions of creating this advantage. The indicated reasons for the competitive advantage are:

- better knowledge of the characteristics of customers and their environment;
- greater knowledge resources compared to competitors;
- the ability to act faster, at lower costs and with a better product offer;
- having such knowledge that is difficult to transfer and therefore obtain by competitors;
- having employees that competing companies do not have;
- speed of assimilation of new knowledge and its application, continuous learning.

The distribution of knowledge in their companies as perceived by the respondents was in most cases uneven (only one respondent stated otherwise). The uneven distribution of knowledge resulted mainly from the diverse experience of employees (e.g. internships in a company), as well as in various areas of specialization. Some types of knowledge, according to the respondents, should be distributed evenly (e.g. knowledge needed by the implementation team or general knowledge about the functioning of the organization), while others should not (specialist knowledge needed by a given employee to perform their tasks). The distribution of knowledge was also related to the organizational structure and responsibility for decisions as well as the need (or lack thereof) for all employees to have similar knowledge (e.g. in accounting firms). Factors influencing the distribution of knowledge in the surveyed organizations are shown in Fig. 7.2. The uneven distribution of knowledge in some way favours the flows of knowledge, as they allow access to knowledge that is missing at a given moment. It can be expected that in the case of an even distribution

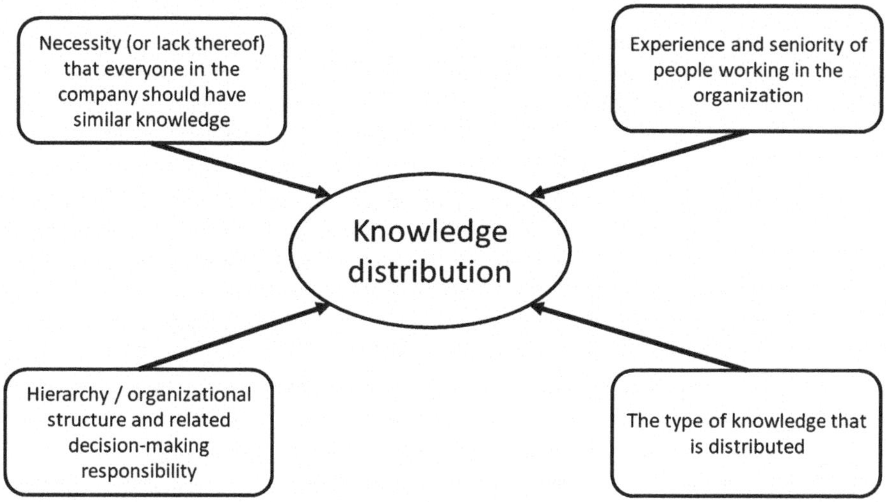

Fig. 7.2 Factors influencing the distribution of knowledge in the surveyed organizations. Source: Author's own illustration

of knowledge in the company, knowledge flows will not be as intense and necessary for the implementation of tasks.

Taking into account the importance of knowledge for KIBS companies, the respondents were asked to indicate the potential threats related to it and the methods of their elimination. This is an important issue, and so far it has not been investigated what knowledge-related threats exist in organizations (not only in the KIBS sector). The most frequently indicated threat was the loss of knowledge related to the employee's departure. It should be remembered that in the case of companies from the SME sector, even a short-term absence of an employee may lead to significant threats resulting from difficulties in replacing the absent resource with another (Durst & Wilhelm, 2011). At the same time, leaving employees often join competing organizations or even set up their own companies to provide similar services. This is due to with high specialization in a given area of knowledge, necessary to perform tasks at KIBS.

Other threats indicated by respondents include: problems in communication and knowledge flow, lack of codified/documented knowledge, knowledge theft (e.g. by competitors), having false knowledge or inability to properly use real knowledge (e.g. in the case of customers), knowledge contradiction acquired from various sources, lack of assimilation of knowledge or even the client's denial of receiving knowledge, aging of knowledge and excess of knowledge. When it comes to ways to eliminate knowledge-related threats, the surveyed companies indicated, inter alia, for such solutions as: verification of knowledge sources, focusing on the knowledge necessary to provide services, confirming receipt of knowledge by the client, sharing knowledge within the KIBS company.

In the area of knowledge management practices applied in the surveyed companies, all companies, without exception, declared building and maintaining employees' knowledge and skills in various forms, including seminars, specialist training courses, knowledge refresher courses or mentoring. This underlines the importance of employees' knowledge for the functioning of KIBS companies, not only with regard to formal qualifications (Miles, 2005), but also in terms of knowledge acquired in complementary forms of education. In terms of willingness to invest in the development of their resources, the surveyed KIBS companies differ from many companies in the SME sector, which are reluctant to make such investments for fear of losing an employee with better competences.

Other popular KM practices among the surveyed companies included:

- acquiring and storing technical knowledge about the market in an electronic repository or in the form of documents;
- using e-mail to share and transfer technical and market knowledge;
- identification and dissemination of internal or external best practices regarding technological, market or operational solutions;
- creating conditions conducive to the exchange of knowledge (e.g. availability of conference rooms, organization of rest areas for employees);
- rewarding (financial or non-financial) employees who share knowledge (with particular emphasis on non-financial forms of reward, e.g. through verbal praise);
- using a community of practice to share knowledge.

The least frequently introduced practices are the implementation of CRM or ERP software and the use of social media to publish and obtain information. While the lack of first practice is not surprising (small businesses rarely use this type of software), the lack of social media does. Social media is becoming an integral part of everyday life in the area of communication and information sharing. They are often free and easy to use, and therefore can provide businesses with a relatively quick and low-cost method of contacting customers. Social media can be particularly beneficial for small- and medium-sized businesses that may not have the financial background or technical expertise required for other, more traditional technology solutions (McCann & Barlow, 2015). A study by Ainin et al. (2015) showed that using Facebook has a strong positive impact on the financial performance of small- and medium-sized businesses. In addition, it was found that the use of Facebook has a positive impact on the non-financial results of SMEs in terms of reducing marketing and customer service costs, improving customer relations and better availability of information (Ainin et al., 2015). The fact that the surveyed companies offering knowledge-intensive business services rarely declared using social media is surprising. The main reasons for this were the perceived lack of need for this type of solution as well as the lack of time. The lack of use of social media may also be related to the lack of the need to actively seek customers, as declared by some respondents.

Knowledge flows were understood by the respondents in a way consistent with the definition proposed. The importance of disseminating knowledge, exchanging/

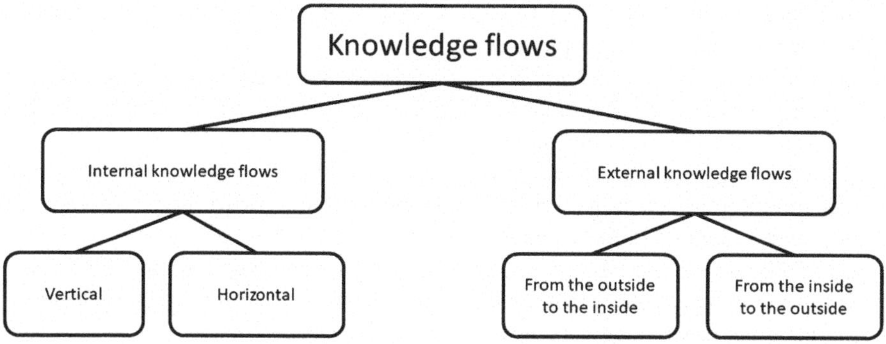

Fig. 7.3 Types of knowledge flows indicated by the respondents. Source: Author's own illustration

transferring knowledge, having it by all those who need it in everyday activities was emphasized. The flows, according to the respondents, took place both within the organization and between the organization and its surroundings. The flow of knowledge was also linked by some respondents with the flow of information, as well as with the experience that is passed on by some employees to others. The flow of knowledge is closely related to the flows of information and data that support the flow of knowledge (Nissen, 2005). The respondents indicated different types of knowledge flows, with the division into internal and external knowledge flows most often indicated. The types of knowledge flows are presented in Fig. 7.3.

The dimensions of knowledge flows identified in the surveyed companies are in accordance with the taxonomy proposed in the first chapter of this book:

- *flow direction*—one-way flows (e.g. from manager to employee) and two-way flows (e.g. from customer to KIBS employee and vice versa);
- *duration of the flow*—synchronous flow (e.g. employee conversations with each other) and asynchronous flow (e.g. knowledge flow by e-mail, development of teaching materials by one employee for others);
- *flow time limits*—urgent flow (e.g. knowledge transfer from one employee to another during a crisis when implementing the service in the client company) and normal flow (e.g. employee training by another employee who participated in the training);
- *flow location*—internal flows of knowledge (e.g. between employees) and external flows of knowledge (e.g. between an employee of KIBS and a training company whose services were used by KIBS);
- *the relationship between the sender and the recipient of knowledge in the flow*—mentoring relationship (e.g. transferring knowledge to younger employees by older employees), coaching relationship (e.g. knowledge flows during professional coaching, related to employee development, or group coaching, during training), peer relationship (e.g. knowledge flow in during the joint exit of

employees "for a beer"), management relationship (e.g. the flow of knowledge from the superior to the employee);
- *the level of flow in the organization*—individual knowledge flow (e.g. between two employees), group/team knowledge flow (e.g. employee training by another employee who has been delegated to training);
- *flow planning*—intended/planned flow (e.g. during daily morning staff meetings and short problem/activity reports), incidental/ad hoc flows (e.g. spontaneous employee meetings in the company kitchen);
- *flow formalism*—formal flows (e.g. transfer of knowledge to the client and confirmation by the client that the knowledge has been acknowledged by signing) and informal flows (e.g. staff meetings "over beer");
- *flow initiating side*—push flow (e.g. flows initiated by employees who have knowledge and want to transfer it to their colleagues—one employee acquires some knowledge and passes it on to others) and pull flow (e.g. knowledge flows initiated by an employee who does not have the knowledge, and he needs it, so I look for this knowledge from others);
- *form of knowledge*—tacit knowledge flow (e.g. transferring knowledge about employee X's experience to employee Y during a conversation), explicit knowledge flow (e.g. knowledge flow about legal acts related to the company's operations), tacit and explicit knowledge flow (knowledge flow during which both forms of knowledge flow or one form of knowledge changes into the other, e.g. an employee who participated in the training provides his colleagues with materials from this training along with his tacit knowledge acquired during the training);
- *knowledge area*—flow of general knowledge (e.g. flow of knowledge on new technologies), flow of business knowledge (e.g. flow of knowledge about customers and their characteristics);
- *the importance of knowledge*—the flow of key knowledge (the flow of knowledge, without which the task cannot be completed—the employee seeks this knowledge to be able to complete the task), the flow of useful knowledge (the flow of knowledge useful at work, but not crucial at a given moment, e.g. knowledge about new regulations legal), the flow of irrelevant knowledge (the flow of knowledge that is not needed, but flows, e.g. knowledge about the private life of employees).

The only type of flows not identified in the study is intergroup/intergroup flow—flow from one group to another or from one team to another. Failure to identify this type of flow may, in part, result from the nature of the surveyed organizations (small- and medium-sized enterprises) and the number of their employees. Small businesses, in particular, may not have enough staff for such a flow to occur. The verified taxonomy of knowledge flows is presented in Fig. 7.4.

Overall, the taxonomy seems to perform well in ordering the dimensions of knowledge flows. In the surveyed companies, it was possible to identify almost all types of knowledge flows that were proposed in the taxonomy. Thanks to the taxonomy, the identification and analysis of knowledge flows in organizations can

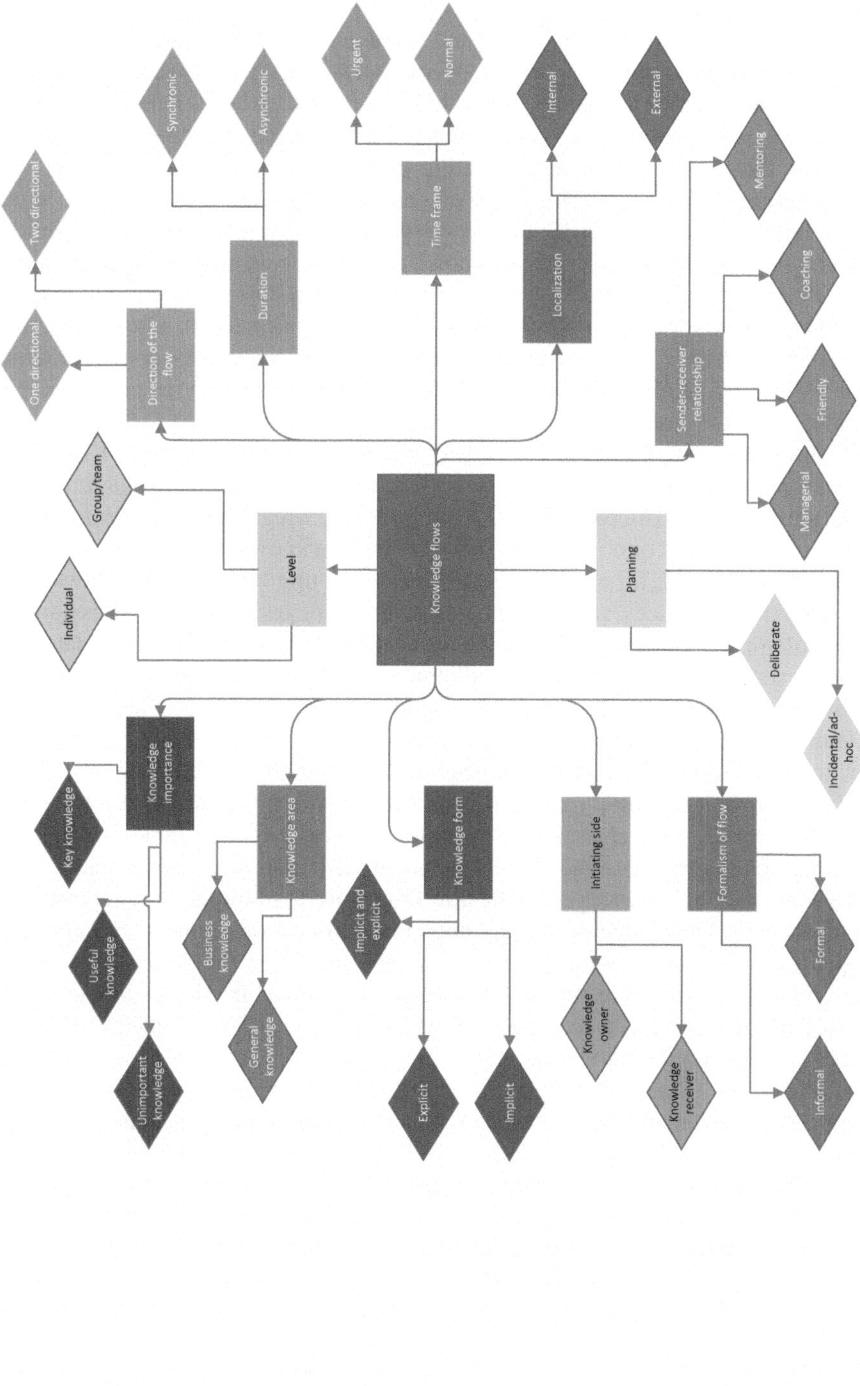

Fig. 7.4 Verified taxonomy of knowledge flows. Source: Author's own illustration

be easier. It should be noted that the flows of knowledge may belong to different categories simultaneously, e.g. the flow may be unidirectional, internal, incidental and informal. Therefore, the proposed dimensions of the knowledge flows are not disjoint.

Among the different types of knowledge flows, flows between KIBS and the client deserve special attention. In the literature, you can find publications emphasizing the importance of close cooperation of the client in creating the KIBS service (Bettencourt et al., 2002; Zieba & Kończyński, 2017). At the same time, other studies show that in the case of creative KIBS services, it may be necessary to even limit the role of the client in creating the service in order to achieve better results (Lehrer et al., 2012). The surveyed companies perceived the importance of knowledge flows between KIBS and the client in different ways. Some companies indicated the need for regular flow of knowledge to the customer and educate him, while others did not see such a need and believed that customers are not able to understand exactly what the service is about and do not need to know it. Some companies emphasized the dependence on the customer's knowledge communicated at the beginning or regularly transferred, without which the service could not be provided. In some cases, two-way knowledge flows were necessary (from client to KIBS and vice versa). The level of intensity of knowledge flows between KIBS and the client may depend on many factors, such as the level of standardization/routine of the KIBS services offered or the type of problem, with which the client reports (whether it is standard or not). In some cases, intensive knowledge flows will be part of the service delivery, for example, in the case of creating "tailor-made" software or some accounting services, when it is necessary to decide on an ongoing basis how to make settlements. The knowledge flows between KIBS and the client are related to problems such as the client's lack of willingness or ability to absorb knowledge, or his low commitment, resulting in dissatisfaction with the service provided.

Another aspect examined was the relationship between knowledge flows and workflows. The flow of knowledge turned out to be naturally related to the workflow in KIBS companies, while in some organizations this flow was sequential (after or before the workflow), and in others—parallel (it occurred at the same time as the execution of tasks (work)). One of the respondents also indicated a close relationship between the flow of knowledge and the flow of resources, resulting from the fact that employees gain knowledge during the implementation of one project, and after its completion, while implementing another project, they pass this knowledge on to their colleagues. This transfer of knowledge is possible thanks to the unusual structure of the respondent's company, in which employees are assigned to various projects and form different teams.

An important element in the operation of KIBS companies is the availability of appropriate knowledge when it is needed. Therefore, one of the questions concerned the issue of knowledge stuck, understood as a situation in which, although someone has knowledge, it is not known who it is and how this knowledge can be obtained. Some companies did not see this type of problem, arguing that their employees are aware of the importance of knowledge and are also jointly responsible for work and do not allow this type of situation. The respondents who pointed to such cases in

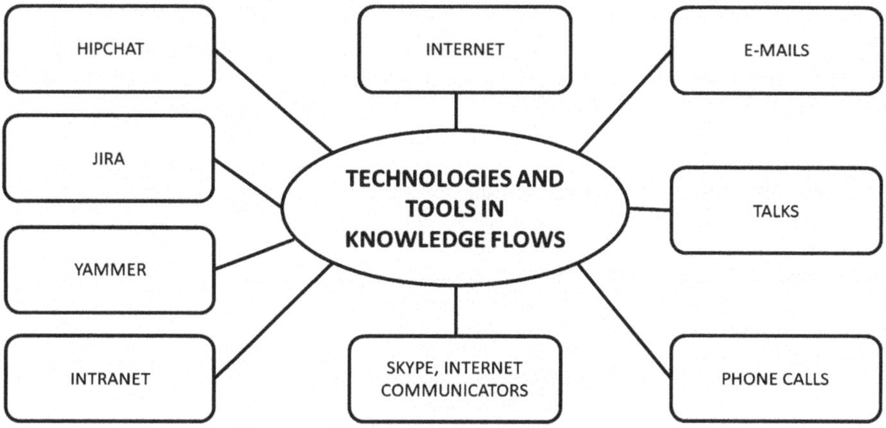

Fig. 7.5 Technologies and tools used for knowledge flows in the surveyed companies. Source: Author's own illustration

their companies were most often able to provide various ways of counteracting such a situation, e.g. by creating an appropriate atmosphere conducive to sharing knowledge or creating a competence matrix and a data bank allowing for quick finding of a person.

Knowledge flows can take place with the use of various types of technologies and tools. Interestingly, the respondents strongly emphasized the importance of direct interaction (mainly "face to face", but also using a telephone or, for example, Skype). The next most popular solution was e-mail. Only individual companies used such special solutions as, for example, Jira, HipChat or Yammer; the majority mentioned only calls, phone calls and e-mails. This may indicate the difficulty in formalizing the flows of knowledge, which are often spontaneous and result from a need that occurs at a given moment. Technologies and tools used for knowledge flows are presented in Fig. 7.5.

Taking into account the importance of knowledge flows for the operation of KIBS companies, the factors influencing various types of flows were also checked. Among the factors influencing the flow of knowledge between KIBS and the client, the respondents included: the appropriate level of detail of the knowledge provided to the client, refining the knowledge (if necessary), articulating the client's needs and problems with the recipient of knowledge (client).

Factors influencing the flow of knowledge within KIBS can be divided into two types based on the respondents' responses: resource factors and technical factors. Resource factors constituted a more numerous category, among which the respondents included, among others trust, limited resources and time, atmosphere in the company, employee characteristics or appropriate motivation. They indicated creating the right atmosphere in the organization by appreciating employees for what they know and rewarding them for sharing knowledge as elements supporting knowledge management in organizations, among others, Zack et al. (2009).

Table 7.1 Factors influencing the flow of knowledge indicated by KIBS companies

Factors influencing knowledge flows between KIBS and the client	Factors influencing the flow of knowledge within KIBS companies	Factors influencing the flow of knowledge between KIBS and the environment
The appropriate level of detail of the knowledge provided to the client Refinement of knowledge (if needed) Articulating needs by the client Problems with the recipient of knowledge (client)	*Resource/human factors* Trust Limited resources and time The atmosphere and characteristics of employees Willingness to cooperate and the willingness to expand knowledge Adequate motivation *Technical factors* Technical limitations Organization of the office	Financial resources Contact with the market

Source: Own elaboration

Appropriate motivation and atmosphere in the company also seem to favour the flow of knowledge. Technical factors included technical limitations related to, for example, the availability of professional portals or the organization of the office in a way that favours the exchange of knowledge (e.g. common space, no doors between rooms). Among the factors influencing the flows between KIBS companies and the environment, the surveyed companies indicated financial resources (thanks to which it is possible, for example, to obtain knowledge during training) and contact with the market, which enables free flow of knowledge. The factors influencing various types of knowledge flows indicated by the respondents are presented in Table 7.1.

Based on the analysis of the research results, an integrative conceptual model of knowledge and its flows was created. This model was created on the basis of the study in small- and medium-sized enterprises offering knowledge-intensive business services. The first part of the model deals with knowledge and related phenomena. Knowledge is presented as an important resource that is of great importance for the functioning of KIBS SMEs. It can be a source of competitive advantage in various ways, for example, by enabling better knowledge of the characteristics of customers and their environment or by enabling faster action compared to the competition. In many cases, knowledge is not evenly distributed within companies, which is sometimes a problem and sometimes it is intentional. Some types of knowledge must be widely available within the organization, while others must not. Sometimes it is necessary that all employees have similar knowledge, and sometimes it is even impossible. There are many risks associated with knowledge, such as loss of knowledge after the employee leaves, communication problems, knowledge theft, having false knowledge or not being able to properly use real knowledge, aging or excess of knowledge. In order to counteract knowledge-related threats, organizations can take various actions, such as knowledge verification, the principle of limited trust in customer relations, knowledge sharing, etc. In the elimination of knowledge-related threats, HR practices can also be helpful, e.g. building and maintaining employees' knowledge and skills, acquiring and storing knowledge, or identifying and disseminating internal or external best practices. Sometimes knowledge becomes stuck, but there are ways to counteract it, such as an atmosphere conducive to sharing knowledge, creating a competence matrix or visual methods of spreading knowledge. Many of the described model elements are related to each other. For example, significant customer knowledge may be evenly distributed across the organization, competitive advantage may be threatened by competitors with theft or HRM practices may contribute to the elimination of knowledge risks.

The second part of the model presents aspects related to knowledge flows in KIBS SMEs. These flows can be of different types, have different dimensions, can be supported by a variety of technologies and tools, and are related to the workflow and flow of resources in organizations. Various types of factors influence the individual flows of knowledge. In the case of flows inside the KIBS company, these may be resource/human factors (e.g. trust, atmosphere or employee characteristics) or technical factors (e.g. organization of an open space office). Important elements in the knowledge flows between KIBS and the customer include competences of the person contacting the client, educating the client, making knowledge available to the client, adjusting to the client's needs, commitment, etc. The knowledge flows between KIBS and the environment depend, among others, on from financial resources (e.g. for training, materials, employee training) or contact with the market. As can be seen in the presented model, knowledge influences the flows of knowledge and related phenomena, and the flows of knowledge influence knowledge and related elements. For example, if a certain knowledge is important and needs to be disseminated within the company, there will be a knowledge flow within KIBS that will make it possible. If there is a risk related to knowledge (e.g. that it will be lost), then by using technology to flow knowledge, this risk can be reduced or eliminated. If the distribution of knowledge in the company depends on experience and decision-making competences, then in relations with the client it is important to delegate the right person from KIBS so that they can acquire the appropriate knowledge from the client and pass it on to him. Such examples could be multiplied. The presented conceptual model integrates elements related to knowledge in KIBS SMEs with elements related to knowledge flows and, according to the author, holistically illustrates the issues related to knowledge and its flows in such organizations (Fig. 7.6).

Although the conceptual model presented above is based on the results of research conducted among KIBS SMEs, it can be cautiously assumed that many of its elements are universal for all types of organizations. Knowledge flows between these companies and the client are a characteristic element of KIBS SMEs. The specificity of the KIBS service, often created in close cooperation with the client, means that knowledge flows between KIBS and the client are intense, which does not always have to be the case with other organizations. The remaining elements of the model seem to be universal in nature, even if the intensity of certain phenomena (e.g., knowledge-related practices, knowledge flows) will be greater in the case of KIBS companies than, for example, in manufacturing companies.

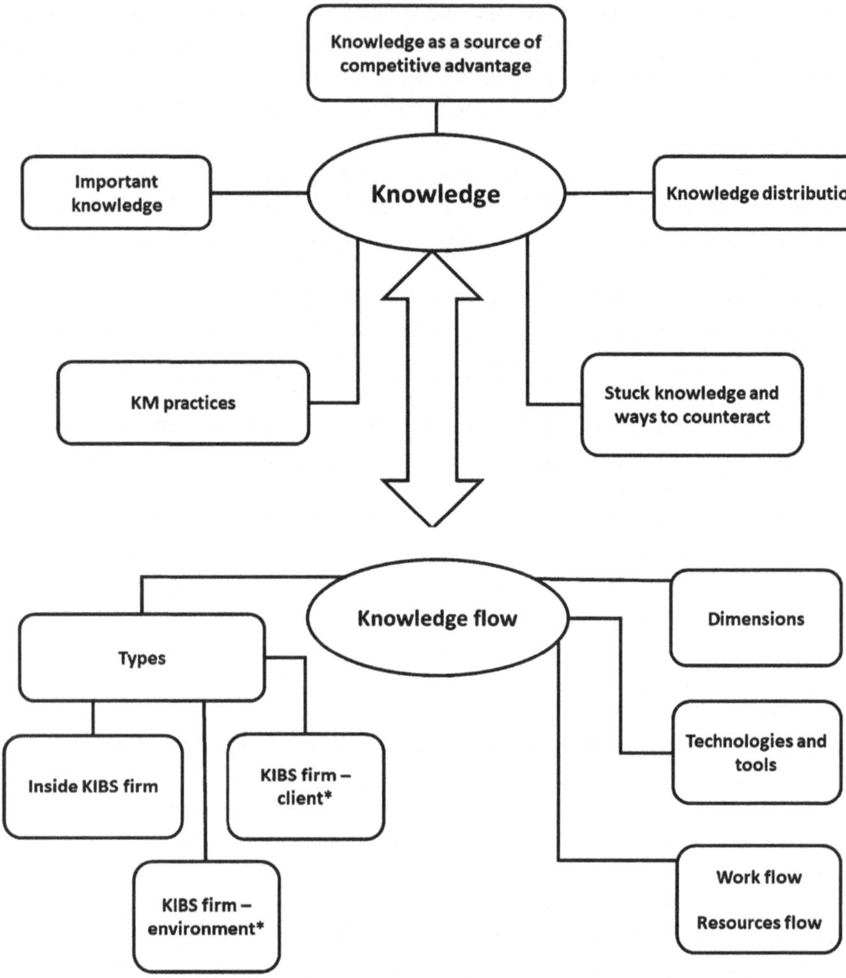

Fig. 7.6 Conceptual model of knowledge and its flows in small- and medium-sized companies offering knowledge-intensive business services. *Bidirectional flows (from KIBS to customer and vice versa and from KIBS to the environment and vice versa). Source: Author's own illustration

Summing up, this chapter presents the research part of the work, starting from the theoretical framework, through the methodological basis and research description, comparing the results of empirical research and discussing them in relation to the existing literature on the subject, and ending with the presentation of the conceptual model of knowledge and its flows. Among the analysed issues there were such as, for example, the way in which knowledge influences the competitive advantage of an organization, knowledge distribution (even/uneven) or practices related to knowledge management introduced in the surveyed entities. An important part of the chapter was the issues related to knowledge flows in companies offering knowledge-intensive business services, including the understanding of flows by respondents, types of knowledge flows, their dimensions, as well as knowledge flows between KIBS companies and their clients, knowledge flows within KIBS companies and between KIBS companies and the environment. The relationship between knowledge flows and work and resource flows, knowledge stuck and how to counteract it, the use of different technologies and tools for knowledge flows, and factors influencing different types of knowledge flows were also examined. In the last part, a conceptual model of knowledge and its flows was created and proposed. All these issues constitute a comprehensive research picture showing the flow of knowledge in the context of companies offering knowledge-intensive business services.

7.2 Conclusions

The main goals of this work included a comprehensive study and explanation of the phenomenon of knowledge flows within small- and medium-sized companies offering knowledge-intensive business services, as well as between them and their broadly understood environment, and to determine the importance of knowledge for this type of companies and the way in which knowledge influences their competitive advantage. In order to achieve these assumptions, the literature was first reviewed and a theoretical framework was created for the concept of knowledge flows. Then, a case study was conducted in 14 selected KIBS SMEs. The collected empirical material allowed for the validation and modification of the created theoretical framework in the context of the studied organizations. The finally formulated theoretical framework is based on the existing knowledge on knowledge management and companies offering knowledge-intensive business services presented in contemporary literature, as well as on the rich material obtained during empirical research. Therefore, it can be concluded that the assumed goals were achieved both in theoretical and empirical terms. The book contributes to the development of the discipline of management sciences by expanding scientific knowledge through the development of the concept of knowledge flows in the KIBS sector, as well as through empirical research and systematization of the existing scientific achievements in the research areas discussed.

The author of this book hopes that the undertaken research effort will prove useful from the perspective of creating the theory of knowledge flows in organizations, and in a broader context—knowledge management. The following can be considered as an original contribution to the development of knowledge management within management sciences field:

- indication (based on an in-depth literature analysis) of a research gap in the area of knowledge, its flows and knowledge management in companies offering knowledge-intensive business services;
- developing the characteristics of knowledge constituting the key resource of KIBS companies and systematizing its features;

- identifying the most important types of knowledge according to the surveyed companies KIBS SMEs and determining how they affect the competitive advantage of these companies;
- determining the ways of perceiving and using knowledge by the surveyed KIBS SMEs;
- developing and verifying a taxonomy of knowledge flows both within the surveyed companies and between these companies and their environment;
- identifying factors influencing different types of knowledge flows;
- deciding what kind of KM practices is introduced in this type of companies, with particular emphasis on the reasons for their introduction or not;
- developing a conceptual model of knowledge and its flows.

Taking into account the location of this work in the area of knowledge management, the impact of the book on the development of this area can be determined. Firstly, the book proposes and verifies the concept of knowledge flows that has not yet been examined from the perspective of the organization and its environment. This concept allows for a better understanding of knowledge management in organizations and complements it. Secondly, a taxonomy of knowledge flows was created and tested, which significantly facilitates their further study and allows their identification in organizations. Third, factors that may affect the flow of knowledge both inside and outside the organization have been identified. Fourth, a conceptual model of knowledge and its flows has been created. Overall, the work presents a broad approach to the concept of knowledge flows along with their characteristics, types and potential determinants. An additional, significant contribution is to fill the research gap concerning an important sector of the economy—the sector of knowledge-intensive business services, which has a significant impact on the development of the knowledge-based economy. This gap included knowledge management in this type of company—an area so far unexplored to a sufficient degree. It should be assumed that the importance of KIBS firms will grow, especially in Europe and the United States, where traditional production is shifting to low-cost countries, as they are an important part of the evolution of the regional and national economy towards more sophisticated service-based systems. KIBS companies can help manufacturing companies stay competitive by producing value-added services related to key processes such as product innovation, design, marketing and technology. KIBS companies are of strategic importance to other organizations as they allow them to participate fully in the knowledge economy as they are an important part of the evolution of the regional and national economy towards more sophisticated service-based systems (Bettiol et al., 2015). Taking this into account, it is additionally justified to better understand the specifics of this type of companies and to examine how they operate and treat their basic resource—knowledge.

Like every work, this one is also not free from limitations. One of the undoubted shortcomings of the study is conducting interviews with one person from a given organization. Admittedly, this approach is used in research in the area of knowledge management (Nunes et al., 2006; Bishop et al., 2008); however, including more people from a given organization in the survey would provide a more complete picture. Taking into account the time and resource constraints, involving a few people from each organization in the study would result in the need to reduce the research sample and obtain less diverse research results. The study could also be enriched by including the clients of KIBS companies in the analysis. This would allow for a more complete picture of the knowledge flows between KIBS companies and customers. In addition, observations from customer companies would be a source of additional knowledge. The problem, however, is the reluctance of KIBS companies to share their clients' data for fear of discouraging them from using their own services. Such a justification for refusing access to customers was found during the investigation. One should also be aware that the study of the KIBS SME sector poses such problems as, for example, a large variety of entities (including both law firms and IT companies, companies offering accounting services, as well as companies providing advertising services), requiring other competences, tools, and finally to approach knowledge. This diversity also overlaps with the great diversity of the SME sector to which the surveyed companies belong. For example, a company with five employees will have different characteristics than an organization with 240 people. When analysing the results of the study, one should be aware that qualitative studies by definition do not offer the possibility of generalizing the results to the entire population.

This work opens up many new research areas. First, the relationship between KIBS and its customer can be viewed and analysed as a kind of network relationship. Knowledge is dynamic and is contained in actors' networks (Spender, 1996). This is one perspective that would be worth exploring in relation to knowledge flows. Secondly, on the basis of the results of the qualitative research, it would be possible to develop a tool for quantitative research, allowing for the analysis of the scale of the presented phenomena (e.g. the occurrence of various types of flows or factors influencing them). Third, knowledge flows could be explored in companies from other sectors as well as other sizes. For example, it can be expected that in larger companies, the flow of knowledge will be more complex or will use more diverse communication channels. The larger the company, the more resources it will use to transfer its knowledge outside (e.g. to clients, contractors, the broadly understood environment, like potential future employees) and also the more internally structured flows of knowledge it will create (e.g. a map of knowledge flows, who, what, when, etc.). Although this book is intended by the author to be a source of knowledge about knowledge, its flows and companies offering knowledge-intensive business services, many threads appearing in the work have not been developed, and others have been deliberately omitted. It resulted from the desire to clearly present the rather complicated matter of the presented issues.

References

Ainin, S., Parveen, F., Moghavvemi, S., Jaafar, N. I., & Mohd Shuib, N. L. (2015). Factors influencing the use of social media by SMEs and its performance outcomes. *Industrial Management & Data Systems, 115*(3), 570–588. https://doi.org/10.1108/IMDS-07-2014-0205

Bettencourt, L. A., Ostrom, A. L., Brown, S. W., & Roundtree, R. I. (2002). Client co-production in knowledge-intensive business services. *California Management Review, 44*(4), 100–128. https://doi.org/10.2307/41166145

Bettiol, M., Di Maria, E., & Grandinetti, R. (2015). Service customisation and standardisation in combinatory knowledge-intensive business services. *International Journal of Knowledge-Based Development, 6*(3), 241. https://doi.org/10.1504/IJKBD.2015.072812

Bishop, J., Bouchlaghem, D., Glass, J., & Matsumoto, I. (2008). Ensuring the effectiveness of a knowledge management initiative. *Journal of Knowledge Management, 12*(4), 16–29. https://doi.org/10.1108/13673270810884228

Corrocher, N., Cusmano, L., & Morrison, A. (2009). Modes of innovation in knowledge-intensive business services evidence from Lombardy. *Journal of Evolutionary Economics, 19*, 173–196. https://doi.org/10.1007/s00191-008-0128-2

Daft, R. L. (2010). Organization theory and design. In *Journal of Chemical Information and Modeling* (10th ed., Vol. 53, Issue 9). South-Western Cengage Learning. https://doi.org/10.1017/CBO9781107415324.004

Durst, S., & Edvardsson, I. R. (2012). Knowledge management in SMEs: a literature review. *Journal of Knowledge Management, 16*(6), 879–903. https://doi.org/10.1108/13673271211276173

Durst, S., & Wilhelm, S. (2011). Knowledge management in practice: insights into a medium-sized enterprise's exposure to knowledge loss. *Prometheus, 29*(1), 23–38. https://doi.org/10.1080/08109028.2011.565693

Koch, A., & Strotmann, H. (2008). Absorptive capacity and innovation in the knowledge intensive business service sector. *Economics of Innovation and New Technology, 17*(6), 511–531. https://doi.org/10.1080/10438590701222987

Lehrer, M., Ordanini, A., DeFillippi, R., & Miozzo, M. (2012). Challenging the orthodoxy of value co-creation theory: A contingent view of co-production in design-intensive business services. *European Management Journal, 30*(6), 499–509. https://doi.org/10.1016/j.emj.2012.07.006

McCann, M., & Barlow, A. (2015). Use and measurement of social media for SMEs. *Journal of Small Business and Enterprise Development, 22*(2), 273–287. https://doi.org/10.1108/EL-01-2014-0022

Miles, I., Kastrinos, N., Flanagan, K., Bilderbeek, R., Den Hertog, P., Huntink, W., & Bouman, M. (1995). Knowledge-intensive business services. Users, carriers and sources of innovation. *A Report to DG13 SPRINT-EIMS, March*, 1–117.

Miles, I. (2005). Knowledge intensive business services: prospects and policies. *Foresight, 7*(6), 39–63. https://doi.org/10.1108/14636680510630939

Muller, E., & Doloreux, D. (2009). What we should know about knowledge-intensive business services. *Technology in Society, 31*(1), 64–72. https://doi.org/10.1016/j.techsoc.2008.10.001

Nissen, M. E. (2005). Dynamic knowledge patterns to inform design: a field study of knowledge stocks and flows in an extreme organization. *Journal of Management Information Systems, 22*(3), 225–263. https://doi.org/10.2753/MIS0742-1222220308

Nunes, B. M., Annansingh, F., Eaglestone, B., & Wakefield, R. (2006). Knowledge management issues in knowledge-intensive SMEs. *Journal of Documentation, 62*(1), 101–119. https://doi.org/10.1108/00220410610642075

Saini, D. S., & Budhwar, P. S. (2008). Managing the human resource in Indian SMEs: The role of indigenous realities. *Journal of World Business, 43*, 417–434. https://doi.org/10.1016/j.jwb.2008.03.004

Spender, J. C. (1996). Making knowledge the basis of a dynamic theory of the firm. *Strategic Management Journal, 17*, 45–62.

Zack, M., McKeen, J., & Singh, S. (2009). Knowledge management and organizational performance: an exploratory analysis. *Journal of Knowledge Management, 13*(6), 392–409. https://doi.org/10.1108/13673270910997088

Zieba, M., & Kończyński, P. (2017). Client co-production in knowledge-intensive business services (KIBS): Case study analysis. *Proceedings of the European Conference on Knowledge Management, ECKM, 2*, 1075–1083.

The manufacturer's authorised representative in the EU is Springer Nature Customer Service Centre GmbH, Europaplatz 3, 69115 Heidelberg, Germany. If you have any concerns regarding our products, please contact ProductSafety@springernature.com

Printed and bound by CPI Group (UK) Ltd, Croydon, CR0 4YY
30/09/2026
02225650-0004